ELECTRICAL SAFETY

ELECTRICAL SAFETY

Rob Zachariason

DELMAR
CENGAGE Learning™

Australia • Brazil • Japan • Korea • Mexico • Singapore • Spain • United Kingdom • United States

**Electrical Safety,
First Edition**
Rob Zachariason

Vice President, Career and
  Professional Editorial:
  Dave Garza

Director of Learning Solutions:
  Sandy Clark

Acquisitions Editor:
  Stacy Masucci

Managing Editor: Larry Main

Senior Product Manager:
  John Fisher

Editorial Assistant:
  Andrea Timpano

Vice President, Career and
  Professional Marketing:
  Jennifer Baker

Marketing Director:
  Deborah Yarnell

Marketing Manager:
  Kathryn Hall

Marketing Coordinator:
  Mark Pierro

Production Director:
  Wendy Troeger

Art Director: David Arsenault

Technology Project Manager:
  Joe Pliss

Project Management:
  PreMediaGlobal

For product information and
technology assistance, contact us at **Cengage Learning
Customer & Sales Support, 1-800-354-9706**

For permission to use material from this text or product,
submit all requests online at **cengage.com/permissions**
Further permissions questions can be emailed to
permissionrequest@cengage.com

Library of Congress Control Number: 2010936496

ISBN-13: 978-1-4354-8185-5

ISBN-10: 1-4354-8185-2

Delmar
5 Maxwell Drive
Clifton Park, NY 12065-2919
USA

Cengage Learning is a leading provider of customized learning solutions with office locations around the globe, including Singapore, the United Kingdom, Australia, Mexico, Brazil, and Japan. Locate your local office at: **international.cengage.com/global**

Cengage Learning products are represented in Canada by Nelson Education, Ltd.

To learn more about Delmar, visit **www.cengage.com/delmar**

Purchase any of our products at your local college store or at our preferred online store **www.cengagebrain.com**

Printed in the USA
2 3 4 5 6   28 27 26 25 24

Contents

CHAPTER 3
Electrical Shock

CHAPTER 4
Arcing Incidents

CHAPTER 5
Working on Energized Equipment

CHAPTER 6
Lockout-Tagout

CHAPTER 7
Personal Protective Equipment

CHAPTER 8
Tool and Equipment Safety

CHAPTER 9
Hazardous Working Environments

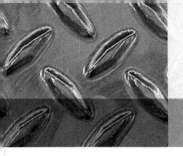

Today's construction jobs are being run at a faster and faster pace; the cost and speed of completion are primary concerns. In trying to meet deadlines and limit costs, safety can sometimes be jeopardized. Rushing, cutting corners, and other sloppy practices can lead to unsafe work environments. It is important to remember that safety is the responsibility of each and every worker on the jobsite. There is nothing more important than the worker's ability to go home safely at the end of each day. In order to ensure one's safety at work, one must understand the potential dangers, follow the safety guidelines, and stay alert.

Intended Use

This text is written for the entry-level electrical worker, electrical students in high schools and two- or four-year college programs, or students in apprenticeship training programs.

There are many hazards out there waiting for the electrical worker, but if a person doesn't know about them they won't be able to protect themselves. The goal of this text is to introduce the reader to the hazards they will encounter while working in the electrical industry. While it is impossible to address every potential scenario, this text introduces the common dangers and stresses the importance of a good attitude toward safety. In addition to identifying potential dangers, this text discusses how to work safely around each particular hazard, and identifies the personal protective equipment that must be worn.

Approach

Construction site safety affects all workers, but each trade looks at it from a slightly different angle as they undertake different tasks. This textbook looks at safety from the viewpoint of the electrical worker, and is written in language that is clear and easy to understand, using images and examples to help the reader relate.

Since electrical hazards aren't the only potential danger an electrical worker will encounter, this text also discusses some of the common nonelectrical hazards present on jobsites, how to work safely around them, and the personal protective equipment that will be required.

Each chapter includes a list of objectives, an introduction, a summary, and review questions. The book begins with an overview of safety hazards to give the reader a general introduction to potential hazards, and get him or her thinking. The text then looks at safety organizations and standards, such as OSHA and NFPA 70E, and the role they play. The importance of OSHA, NFPA 70E, and employer safety programs are stressed throughout the book. The remainder of the text goes into detail about the various types of hazards and how to work safely with and around them.

Supplement Package

The Instructor Resource is geared to provide instructors with all the tools they need in one convenient package. Instructors will find that this resource provides them with a far-reaching teaching partner that includes the following:

- PowerPoint® slides for each chapter that reinforce key points and feature illustrations and photos from the book. The PowerPoint also allows instructors to tailor the course to meet the needs of their individual class.
- A computerized Test Bank in Exam View format, which allows test customization for evaluating student comprehension of noteworthy concepts.
- An electronic Instructors Manual with the answers to the textbook review questions.
- An image library that contains nearly all the images in the book and can be used to enhance the PowerPoint presentations or create transparencies and handouts.

Instructor Resource ISBN – 1-4354-8186-0

About the Author

Rob Zachariason is an instructor at Minnesota State Community and Technical College and an instructor for the Joint Apprenticeship and Training Committee. Rob is a member of the International Brotherhood of Electrical Workers, the International Association of Electrical Inspectors, and the National Education Association, and he holds Master Electrician licenses in North Dakota and Minnesota. Rob worked as an electrician for ten years before becoming a full-time instructor.

Acknowledgements

I would like to thank my wife Brandi and my children Lauren, Kate, and Julia for their understanding and support while I was working on this text.

I would also like to thank the entire Delmar staff for their hard work and support, but in particular Stacy Masucci and John Fischer.

Thank you to the following people for their assistance in acquiring images for the text:

Amber Scott – Versalift (Time Manufacturing)

Carrie Schank – Salisbury by Honeywell

Lisa Alsdorf – Miller by Sperion

Erica Aratari – Carhartt

Mike Marquardt – Red Wing Shoes

Fred Blosser – CDC/NIOSH

The author and publisher would like to thank the following reviewers for their contributions:

Mark Burch
Honolulu Community College
Honolulu, HI

Marvin Moak
Hinds Community College
Raymond, MS

Jim Richardson
Lee College
Baytown, TX

CHAPTER 1

Safety Overview

Objectives

- Describe the purpose of construction safety
- Identify who is responsible for safety
- Identify dangers on a jobsite
- Describe the dangers of horseplay as it relates to a construction site
- Describe the dangers of alcohol and drug use as it relates to a construction site

Introduction

The construction industry provides many extremely rewarding careers, where people can work with their hands and have the satisfaction of witnessing the fruit of their labor. This type of career involves working in industrial settings and on construction sites, which can be dangerous places. (Figure 1-1) To ensure your safety, and the safety of others, there are three key things you must do: stay alert, understand the hazards involved, and follow all safety guidelines.

Safety is the responsibility of every person involved with a construction site, from the electrical contractor to the apprentice. The contractor plays an important role by ensuring that employees are following all safety guidelines, a role which includes having and enforcing a safety plan. Every worker is responsible for looking out for hazards that may endanger themselves or another person on the job, and for following all safety guidelines. People new to construction, such as apprentices, must be instructed about the hazards involved, as they may not know what dangers exist. Ultimately, the final responsibility for

Figure 1-1 Construction sites can be dangerous places. They often involve heavy equipment being operated, falling objects, loud noises, fall hazards, and many other dangers. *Courtesy Delmar/Cengage Learning*

working safely falls on your shoulders; you have control over what you do and therefore have the final say on what is safe or what is not.

Safety is a mindset that must be instilled from the start. A bad habit is hard to break, so it is important to learn good safety habits from the very beginning. It is also important to set a good example. New people in the trade will look to experienced workers for guidance. By being safety-conscious and leading by example, new workers will learn the proper safety procedures right from the start. For example, wearing safety glasses and requiring them for all workers will become a good habit that may save your eyesight, or that of others. Once a person has become accustomed to wearing glasses, they will feel as if they are missing something without them. However, if a person is not used to them and will only wear them when they think there may be a hazard, they will tend to forget them and be uncomfortable with them on, and the unexpected may someday take their eyesight. It is also important to show a positive attitude towards safety. By complaining about safety regulations and requirements you will create a negative attitude in others. Understanding the need for safety—and embracing it—will catch on with other workers on the job, and may save a person's life.

Construction Awareness

Awareness of your surroundings is extremely important. You should constantly be thinking about potential hazards around you and considering what dangers may be present on the job. (Figure 1-2) The following are just a few examples to bear in mind:

- Potential hazards you are working on or around (power tools, energized equipment, etc.);
- Other people using power tools, and their possible dangers;
- People working above you (falling or dropping tools or materials);
- Escape routes;
- Pinch points;
- Railings and openings that could create a fall hazard;
- Items that could be electrically energized;
- Location of the First Aid kit(s).

Figure 1-2 While working on a construction site, a worker must always be aware of his or her surroundings to prevent an accident (in this case, a fall hazard). *Courtesy Delmar/Cengage Learning*

As time goes on, jobs are being completed at a faster pace: everything is expected to have been done yesterday. This can cause people to try to cut corners in order to finish tasks faster. Many times these corners cut are at the expense of safety, and ultimately lead to accidents. Rushing a task tends to make a person careless and may end in an accident. Trying to save a few minutes by not properly securing a ladder; not properly tying off; not taking the time to follow proper lockout-tagout procedures (Figure 1-3); not using the proper tool for the job; or working with something energized, to give just a few examples, could ultimately end up causing an injury that may end your career or your life.

I have personally witnessed several injuries on the job, and right before they happened I thought to myself: "that isn't a good idea," or "I shouldn't be doing this . . ." If you feel uneasy about what you are about to do, your mind is telling you not to do it. If you are not sure if something is safe, ask someone who does know or can find out.

Many times, the last thing a person thinks before an accident is "this isn't a good idea, because _____ could happen"; and then it

Figure 1-3 Locking out and tagging out equipment will notify people that the equipment has been intentionally shut off and will prevent the equipment from being turned back on. *Courtesy Delmar/Cengage Learning*

does. If you have any doubt about what you are about to do, then don't do it. The following are a few examples of questions that should raise a red flag that what you are about to do is unsafe.

- Is that asbestos?
- Is that a confined space?
- Is this circuit still energized?
- What if this drill slips?
- Should I be wearing a harness?
- If my screwdriver slips, could there be an arc flash?
- What type of personal protective gear should I be wearing?
- Will the workers above me drop something?

Construction Hazards

There are many hazards the electrical worker will come across, not all of which are electrically-related. Falls, confined spaces, asbestos, moving machinery, loud noises, environmental dangers, electrical hazards, and many others all need to be on the mind of the safety-conscious electrician. You should be constantly evaluating your working environment and looking for any potential dangers. This section of the chapter will give a brief overview of the many dangers associated with working in the electrical industry, dangers which will be covered in greater detail in later chapters. This is by no means a complete list of the hazards encountered by an electrician on a daily basis, but a sample that will help begin the process of understanding construction safety.

Hazards of Electricity

There are many hazards an electrician will deal with that are associated with electricity. Most people think only of electrical shock, but there are others, all of which can lead to permanent injury or death.

Electrical Shock

Electrical shocks can vary from being a slight tingle to a shock resulting in death. The severity of a shock is determined by three key factors: the amount of current, the path of the current, and the amount of time the current flows. Always turn the power off before working on a circuit. The only time it may be permitted to work on something "hot" is if this is necessary for taking measurements, or if turning it off presents a greater hazard. If a piece of equipment must be worked on while energized, the proper safety procedure must be followed and the appropriate personal protective gear must be worn. (Figure 1-4)

Arcing Incident

An arcing incident consists of an arc flash and arc blast. (Figure 1-5) Depending on several variables, it could be a small incident (for example, two wires shorting together in a switch box), or it can be very large and rip apart equipment. When conductors or a piece of equipment have a fault—which could be a short circuit or ground fault—a large amount of current flows, causing an explosion which can vaporize copper instantly. This vaporized copper is hotter than the surface of the sun; the shock wave is similar to that of dynamite; and the noise level is damaging to the ears. Understanding the hazard and wearing

Figure 1-4 This worker is wearing voltage-rated gloves with leather protectors to prevent an electrical shock while taking voltage measurements in a meter socket. *Courtesy Delmar/Cengage Learning*

Figure 1-5 Workers exposed to an arc flash are subjected to extremely high temperatures, flying shrapnel, loud noises, blinding light, and very powerful shockwaves. *Courtesy Delmar/Cengage Learning*

the proper personal protective gear can minimize harm in the case of an arcing incident.

Falls (resulting from electrical shock or witnessing an arcing incident)

Electricians often work on ladders or scaffolding. Receiving a shock or witnessing an arcing incident can cause a person to lose their balance and fall.

Jerk Reaction (resulting from electrical shock or from witnessing an arcing incident)

When a person receives an electrical shock or witnesses an arcing incident, their reflex is to jerk and move away. This may cause a person to move their hand into a sharp object or moving machinery, inadvertently hit a start or stop button, or throw an object into another worker or piece of equipment.

General Construction Hazards

In addition to the electrical hazards found on a construction site, there are many hazards that are not electrical. The following is an overview of some of the general construction hazards that will be covered in greater detail in later chapters.

Falls

Construction workers work from elevated platforms and ladders on a daily basis. Injuries from falls are some of the most common on a construction site. Ladder safety, scaffold safety, and using a harness that is properly tied off are essential parts of staying safe when working above ground level. (Figure 1-6)

There are often openings in the floor during the construction process for elevators, stairwells, ventilation ducts, etc. A person must have a constant awareness of them and keep a safe distance.

Confined Spaces

Many people don't understand what is classified as a confined space, or realize the dangers. A few examples of confined spaces are: manholes, the inside of tanks or silos, underground vaults, tunnels in buildings,

Figure 1-6 Scaffolding and other elevated platforms used on construction sites create a fall hazard. Working in these situations requires a worker to have had training on the proper working procedures and wear the appropriate personal protective equipment. *Courtesy Delmar/Cengage Learning*

and trenches. A confined space may contain hazardous gases, have the potential for a cave-in, or lack oxygen. It is important to know and follow the proper procedures when entering a confined space.

Moving Equipment and Machinery

Construction sites typically have equipment such as forklifts, scissor-lifts, earth-moving equipment, trenchers, cranes, etc. It is important that the equipment operator is looking out for people and objects, but it is also important for everybody else on the jobsite to watch out for the equipment.

Industrial jobs will have all the equipment mentioned above, but may also have machinery and equipment which are part of the process of the plant and that must be carefully worked on or around. (Figure 1-7) It is important that workers understand the hazards of the process and maintain not only their personal distance, but also a safe distance with tools and materials. Foreign materials such as a piece of conduit accidentally getting tangled in equipment or even

Figure 1-7 Equipment and machinery will often have moving parts that can cause serious injury if an object or part of the body comes into contact with it. *Courtesy Delmar/Cengage Learning*

pushing a stop button can have devastating effects. Before working on equipment or machinery it is extremely important that all power has been removed and that all stored energy has been released.

Loud Noises

Construction sites and manufacturing plants are noisy places. From pounding nails to operating a hammer drill, a lot of the work an electrician does will create noise that can cause permanent hearing damage. You must also consider the noise created by other workers on a job site or by machinery in an industrial process. There isn't any realistic way to completely eliminate the noise on a jobsite, but by wearing the proper ear protection you can prevent hearing loss.

Flying or Falling Objects

Power tools, hammering, and machinery can all cause objects to be thrown at very high speeds. Safety glasses should be worn at all times, as the flying object may come from you, somebody working near you, or a machine (for example, the saw shown in figure 1-7).

If equipment or people are working above you there is a possibility that a tool or piece of material may fall and hit you on the head. Hard hats are used to help prevent head injuries from falling objects. They may also save your head from a bruise or cut if, for example, you are working in the ceiling and turn into a piece of steel in the rafters, or stand up and hit a metal edge.

Environmental Dangers

Electricians will occasionally have to work in extreme conditions. Winter work may be in sub-zero temperatures, while summer work may take place in temperatures well over 100 degrees. Not wearing the proper attire for the situation can be damaging to the body or even life-threatening. Proper nutrition and hydration also play an important role.

Inhaling Particles or Vapors

Construction is often extremely dusty, especially if it is a remodel. Particles such as asbestos or concrete dust can be hazardous to your health. It is important to use the proper procedures to prevent the inhalation of these items. Simply using a mask may not provide the necessary protection. NOTE: disruption of asbestos must be avoided and removal should only be done by trained professionals.

Toxic vapors and gases are often present on a construction site. A few examples are: PVC glue; fumes from sprayed polyurethane insulation; and the exhaust from portable propane or other types of heaters, propane or diesel lifts, and skid-steer loaders. (Figure 1-8) Proper ventilation is essential to prevent hazardous conditions.

Cuts

Most construction workers will sustain minor cuts and injuries from time to time. Even minor cuts can slow your work down and make things difficult. Some injuries and deep cuts will have lasting effects even after they have healed. For example, a deep cut in a finger will be sensitive to touch and cold long after it has healed. (See Figure 1-9) I have a finger that sustained a significant cut ten years ago, which still

Figure 1-8 Portable heaters that are often used on a construction site can create a hazard if there isn't enough fresh air or they aren't used properly. *Courtesy Delmar/Cengage Learning*

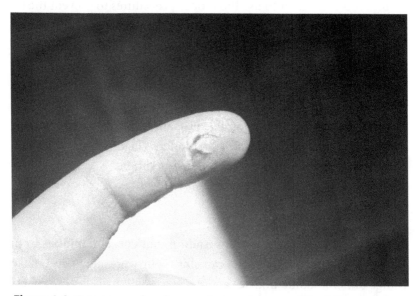

Figure 1-9 Cuts on your hand can make your job very difficult if you work with your hands. *Courtesy Delmar/Cengage Learning*

hurts when bumped and aches when exposed to cold weather. It is not something that is life-threatening, but is a nuisance when you work with your hands.

Working with hand and power tools can very quickly and easily cut or injure a worker. Remember to avoid rushing and take the necessary precautions to avoid injuries from tools. Many of the materials installed by electricians and the other construction trades have extremely sharp edges. A person must always be on the lookout for objects or scenarios which could cause injuries, and do what is necessary to avoid them. It is important to remember that if you feel uneasy about what you are doing, an injury is probably right around the corner. Stop, and do what is necessary to ease your mind and make the task safe. Wearing gloves will help avoid some of the minor cuts and injuries to workers' hands. (Figure 1-10) It is important to have gloves that fit well and allow good finger movement and dexterity.

In the case of a deep cut or injury, seek medical attention immediately. Waiting several hours before getting treatment may cause

Figure 1-10 Work gloves can help protect a worker's hands from minor cuts and injuries. *Courtesy Delmar/Cengage Learning*

the injury to worsen, or limit how a doctor can treat it. For example, if a cut is several hours old, doctors will often not be able to stitch it up. It is also important to let the doctor know that the injury happened at work, so you won't be held personally responsible for the bill and any related treatments which may be necessary at a later date.

Jewelry

Jewelry can create a serious safety hazard when worn at work. Earrings, necklaces, and rings can get caught on or stuck in objects. Many construction workers wearing rings have caused serious injury to their finger due to getting them caught on objects such as the top of a ladder. (Figure 1-11) Some have even torn their finger right off.

Another danger associated with wearing jewelry at work is the possibility of it becoming energized. Rings, necklaces, etc. that come into contact with an energized component can cause electrical shock and may become extremely hot, resulting in severe burns. The only way to avoid these dangers is to leave all jewelry at home.

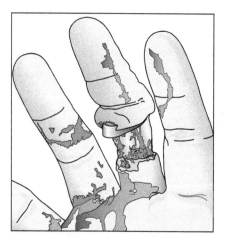

Figure 1-11 Rings shouldn't be worn at work, as they can cause a serious injury if they get caught on something or become energized. *Courtesy Delmar/Cengage Learning*

Horseplay

It is important to have a good relationship with your fellow workers and enjoy working on the job, but there is a time and place for "messing around." Something that starts out innocently, like tossing small objects at someone, pushing, or trying to scare or startle someone, can lead to serious trouble.

A few examples of innocent jokes that can lead to dangerous incidents:

- Throwing an object such as a wire nut, strap, etc. This can startle a person, causing them to cut themselves, move their hand into moving machinery, or look up just as the item is tossed and get hit in the face or eye.
- Pushing or nudging. Even a slight push can cause a person to lose their balance and fall down a flight of stairs, off a building, or onto an object causing injury.
- Pounding on a wall. It may seem funny to use a hammer to hit the opposite side of the ceiling, wall, or floor where someone else is working in order to startle them. The problem is that you don't know what that person is working on. You could cause someone about to hammer something to hit their finger; someone about to cut into a wire to accidently cut themselves; or someone to slip while working in an electrical panel and cause an arc blast.

The little bit of humor and amusement from goofing around isn't worth the potential hazard that comes along with it. Save the messing around for when you aren't at work.

Alcohol and Drugs

There is no place on the jobsite for alcohol and illegal drugs. People with alcohol or illegal drugs in their system create a hazard to themselves and others. They tend to take more risks, have slower reflexes, lack the ability to make good decisions, and have lower productivity. Many employers implement drug testing and have a written no-tolerance drug policy, whereby a person under the influence of illegal drugs or alcohol at work will be immediately terminated.

Hangovers and lack of sleep create hazards similar to those caused by alcohol or illegal drug use. A person who comes into work with a hangover will often become dizzy on a ladder or scaffold, have slower reflexes, lack the ability to make good decisions, and have lower productivity. Studies have shown that lack of sleep can have some of the same effects as drug and alcohol abuse; most adults need seven to eight hours of sleep each day.

There are also a variety of legal drugs that may present a hazard. Many prescription and non-prescription drugs create side effects that may impair workers' abilities. Always read the label of the drug you are taking to see if it will have any adverse effects that may create a hazard. Drugs that may not create any problems on their own can interact when combined and cause a person to become impaired. If taking more than one prescription or non-prescription drug, be sure to check with your doctor or pharmacist about any possible drug interactions that could impair your abilities.

There is also a potential hazard for people who fail to take medication for a medical condition. I was on the job with a person who was diabetic and had forgotten his medication: it was an alarming experience. Most of us on the job didn't know this person was diabetic, or what symptoms to look for. He seemed fine when we first started work, but later in the morning I noticed him lose his balance and fall into a wall. I asked him if he was OK, and he just made a joke about the night before. By the time we sat down for lunch he was unable to communicate clearly. After frantically searching his toolbox for his medication, he got up and tried to walk, but stumbled and fell. One of his co-workers did know he was diabetic, recognized the symptoms, and knew what to do. Luckily this worker wasn't seriously injured, and he returned to work the next day, but the outcome could have been very different. He could have fallen from a scaffold or roof, or stumbled into an open electrical panel or moving piece of machinery; the number of ways he could have been injured or killed are countless. Always keep a lookout for a person who may be having trouble, and don't take it lightly.

Summary

- Construction sites can be a dangerous place to work. Safety is the responsibility of every person on the site, from the electrical contractor to the apprentice. Remember that you have the final say when it comes to your safety.
- Awareness of your surroundings, including other workers, is an important part of construction safety. You should always be surveying the situation and looking for potential dangers.
- There are several potential hazards on a jobsite that are related to electricity:
  - Electrical Shock;
  - Arcing Incident (arc flash/arc blast);
  - Fall (due to a shock or arcing incident);
  - Jerk reaction (due to a shock or arcing incident).
- There are many potential hazards on a jobsite that are not electrically-related:
  - Falls;
  - Confined spaces;
  - Moving equipment and machinery;
  - Loud noises;
  - Flying or falling objects;
  - Environmental dangers;
  - Inhalation of particles or vapors.
- Cuts can be avoided at work by wearing gloves, using caution, and being aware of nearby dangers.
- Jewelry should not be worn at work due to the danger of it becoming electrically energized or caught on an object.
- There is no place for horseplay on a job site. Innocent goofing around may lead to a person sustaining a serious injury or fatal accident.
- People with alcohol or illegal drugs in their system create a hazard to themselves as well as others. People using alcohol or drugs on the job tend to take more risks, have slower reflexes, and lose

the ability to make good decisions. Lack of sleep or a hangover can create symptoms similar to those of a person under the influence of alcohol or drugs.

- By understanding the hazards involved, following safety guidelines, and staying alert you can make the jobsite a safer place.

Review Questions

1. Who is the final decision-maker when it comes to safety?
2. Why is it important to follow proper safety procedures from the very beginning?
3. List three things a person must do to ensure their safety on the jobsite.
4. List three jobsite hazards that are related to electricity.
5. List six jobsite hazards that are not related to electricity.
6. What should a person do before taking prescription drugs at work?
7. Why is horseplay dangerous on a construction site?
8. What are two possible injuries risked by wearing a ring at work?

2

Safety Organizations, Standards, and Certifications

Objectives

- Describe the purpose of the Occupational Safety and Health Administration

- Understand employer responsibilities as required by OSHA

- Understand employee responsibilities as required by OSHA

- Describe the purpose of the NFPA 70E (Standard for Electrical Safety in the Workplace)

- Describe the purpose of the National Institute for Occupational Safety and Health (NIOSH)

- Understand the need for First Aid and CPR training

Introduction

This chapter will give an introduction to various organizations and standards which are intended to help protect the worker. Education about the hazards involved and an understanding of safety requirements is imperative for worker safety. Having safety organizations and standards cannot ensure worker safety if we as employers and employees aren't willing to learn and follow the requirements.

Occupational Safety and Health Administration (OSHA)

The Occupational Safety and Health Administration (OSHA) is a federal organization whose purpose is to ensure the safety of people in the workplace. (Figure 2-1) Since its inception, workplace injuries and fatalities have been dramatically reduced. OSHA has standards that employers and workers are required to follow and that are in place to keep workers safe.

History of OSHA

Prior to 1970 there weren't any comprehensive provisions in place to protect people in the workplace. There were a staggering number of worker deaths, work-related disabilities, and new cases of occupational diseases. In 1970, Congress passed the Occupational Safety and Health Act, which was signed into law by President Nixon. The intention of the act was to ensure that men and women working in the United States enjoy safe working conditions.

**Occupational Safety
and Health Administration**
www.osha.gov

Figure 2-1 OSHA is a government organization dedicated to keeping people safe at work. *Courtesy Occupational Safety and Health Administration*

Under the Occupational Safety and Health Act of 1970, the Occupational Safety and Health Administration (OSHA) was created within the Department of Labor. The purpose of OSHA is to:

- Encourage employers and employees to reduce workplace hazards and to implement new, or improve existing, safety and health programs;
- Provide for research in occupational safety and health to develop innovative ways of dealing with occupational safety and health problems;
- Establish "separate but dependent responsibilities and rights" for employers and employees to further the achievement of better safety and health conditions;
- Maintain a reporting and record-keeping system to monitor job-related injuries and illnesses;
- Establish training programs to increase the number and competence of occupational safety and health personnel;
- Develop mandatory job safety and health standards and enforce them effectively;
- Provide for the development, analysis, evaluation, and approval of state occupational safety and health programs.

(Above taken from the US Department of Labor [All About OSHA])

Although the act is broad and intended to cover all workers, there are a few people who are not covered. People who are self-employed, farms employing only immediate family members, and people working under conditions regulated by other federal agencies or federal statutes are not covered under the act.

OSHA Enforcement

OSHA has the authority to perform workplace inspections on any establishment that is covered under the Occupational Safety and Health Act. An OSHA compliance officer can perform an inspection without any advance notice. The officer does, however, have to present the appropriate credentials before performing the inspection. If an employer does not permit, or interferes with, an inspection, legal action may be taken.

There aren't enough officers or hours in the day for every job to have an OSHA inspection. There are several scenarios which may trigger an OSHA inspection: catastrophic accidents, employee complaints, high-hazard industries, and follow-up inspections. The inspections are categorized according to their priority level as shown in Figure 2-2.

Employers who are found to not be in compliance with OSHA guidelines are cited and may receive a fine, which can amount to several thousand dollars. It all depends on how dangerous the offence is, and whether it is a repeat offence. The employer will have to correct the deficiency within a certain period of time.

OSHA has a 24-hour Emergency Service Hotline (1-800-321-OSHA) for contacting the agency about imminent dangers on the job. The call can be made anonymously for the protection of the caller.

OSHA Inspection Priority

Highest Priority		
Imminent Danger	This is a situation which may result in serious injury or death and must be dealt with immediately and is given top priority.
Catastrophes and Fatal Accidents	Investigations that are to determine if any OSHA standards were violated and to help prevent future accidents.
Employee Complaints	Employees call about unsafe conditions that are an OSHA violation.
Programmed High-Hazard Situations	Inspecting industries and locations which are known to have high risk of hazards.
Follow Up Inspections	Ensure corrections have been made
Lowest Priority		

Figure 2-2 OSHA makes situations with imminent danger their top priority.
Courtesy Delmar/Cengage Learning

Employer Responsibilities

Employers must comply with all OSHA standards and regulations, and they bear the responsibility of providing a workplace that is free from hazards that may cause serious injury or death to their employees. Compliance includes understanding OSHA's requirements and having an effective safety program.

OSHA requires employers to provide the necessary training for their employees so they understand the hazards involved and the necessary steps to ensure that a job is done safely. OSHA also states that if a person hasn't performed a task in one year, the employee must be retrained before performing the task again. Employers have the responsibility to verify, through supervision and inspections, that employees are complying with all safety-related work practices.

Employee Responsibilities

Although the employer has a large role, final responsibility lies with employees, who must comply with all OSHA rules and regulations as well as any safety requirements implemented by the employer. Employees must wear the proper personal protective equipment, use tools appropriately, and report job-related injuries to the employer immediately. It is up to us as employees to make a conscious effort to work safely.

OSHA Classes

Many jobsites will require a person to have taken a 10- or even 30-hour OSHA course. The courses are given by a qualified instructor, and at the end participants receive a card indicating that they have successfully completed the course. (Figure 2-3) Although the card doesn't have an expiration date, many jobs require the course to have been taken within the past one to three years. Every person working on a construction site should have taken an OSHA 10-hour course.

Figure 2-3 After completion of an OSHA 10-hour class a card will be issued. *Courtesy Delmar/Cengage Learning*

There are some companies which offer the OSHA 10- and 30-hour courses online. If you are planning to take an online OSHA course to acquire a card for a particular jobsite, be sure to verify that they will accept an online version of the course. Some jobsites require face-to-face OSHA classes and will not accept OSHA cards that were earned online.

NFPA 70E (Standard for Electrical Safety in the Workplace)

OSHA lays down requirements for electrical safety on the job, but doesn't give specific details about how to comply with them. For example, OSHA requires that a person must wear the appropriate personal protective gear to guard against an arcing incident, but doesn't explain what personal protective gear will be required.

The NFPA 70E (Standard for Electrical Safety in the Workplace) tells us how to protect workers from electrical hazards. (Figure 2-4) For the example above, the NFPA 70E will indicate the level of personal protective gear that is required to provide protection in the case of an arcing incident.

Figure 2-4 The NFPA 70E is the Standard for Electrical Safety in the Workplace. *Courtesy National Fire Protection Agency*

History of NFPA 70E

In its early stages, OSHA attempted to use the National Electrical Code (NEC) as its standard for electrical safety, but it was soon realized that the NEC wasn't suitable for workplace safety. The National Electrical Code is intended to provide an installation free from electrical hazards, but doesn't really address the safety of those installing or working on electrical systems. OSHA needed a document pertaining to the safety of workers installing or working on electrical systems that not only met OSHA's requirements but was consistent with the National Electrical Code. In 1976 OSHA requested that the National Fire Protection Agency (NFPA) create a consensus electrical safety standard.

In 1979, the National Fire Protection Agency released the first recognized electrical safety standard, the NFPA 70E, which was named *Electrical Safety Requirements for Employee Workplaces*. The first version of the NFPA 70E was divided into four parts. (See Figure 2-5)

It was decided that each part was independent and could be published separately. The first edition of the NFPA 70E, published in 1979, only included Part I. The second edition was published in 1981 and included Part I as well as a new Part II. The third edition published in 1983 included Parts I and II as well as a new Part III. Part IV wasn't added until the sixth edition in 2000.

The most current edition is the eighth, published in 2009. Over the years there have been many changes: safety requirements have been updated, and graphs and charts have been added. The name was changed from *Electrical Safety Requirements for Employee Workplaces* to *Standard for Electrical Safety in the Workplace*. The parts were renamed as chapters and reorganized. The chapter that contained "Installation Safety Requirements" was removed, as it duplicated information found in the National Electrical Code. The current edition of the NFPA 70E includes three chapters. (See Figure 2-6)

The NFPA 70E is a consensus standard, which means that it has been written, and is revised by, groups with various interests. The document is written so that it is adoptable, but it is not required by OSHA that it be adopted. The NFPA 70E is written using the same format as the National Electrical Code, making it easy to navigate for those familiar with the NEC.

1979 NFPA 70E "Electrical Safety Requirements for Employee Workplaces"	
Part 1	Installation Safety Requirements
Part 2	Safety Related Work Practices
Part 3	Safety Related Maintenance Requirements
Part 4	Safety Requirements for Special Equipment

Figure 2-5 The parts of the 1979 NFPA 70E. *Courtesy Delmar/Cengage Learning*

2009 NFPA 70E "Standard for Electrical Safety in the Workplace"	
Chapter 1	Safaty-Related Work Practices
Chapter 2	Safety Related Maintenance Requirements
Chapter 3	Safety Requirements for Special Equipment

Figure 2-6 The parts of the 2009 NFPA 70E. *Courtesy Delmar/Cengage Learning*

Purpose of NFPA 70E

The NFPA 70E is intended to provide a workplace that is safe from hazards that may arise from the use of electricity. An employer may adopt the NFPA 70E and follow its requirements, but it is not a re-placement for the employer's safety program. Remember that although electrical safety plays a major role in what we do on the job, it isn't the only danger we will encounter. An employer safety program has to in-clude an electrical safety program as well as all other aspects of safety on the job.

Many electricians have never taken the time to study the NFPA 70E, and some don't even know it exists. The NFPA 70E is a standard that every electrician should know, as following its rules could some-day save your life.

To help users understand and apply the NFPA 70E, the National Fire Protection Agency publishes the *NFPA 70E Handbook*. The hand-book contains the entire NFPA 70E, as well as explanations, pictures, diagrams, etc. (Figure 2-7)

National Institute for Occupational Safety and Health (NIOSH)

The National Institute for Occupational Safety and Health is an agency under the US Department of Labor. (Figure 2-8) It was created by the Occupational Safety and Health Act along with OSHA, and its pur-pose is to help assure that working conditions for men and women are safe and healthful. To this end, the Institute conducts research and

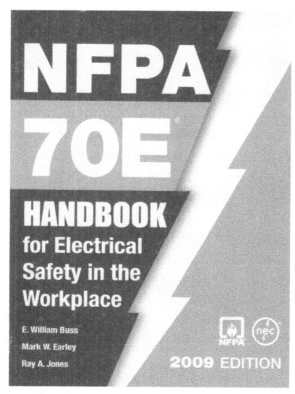

Figure 2-7 The *NFPA 70E Handbook* contains the NFPA 70E as well as pictures and explanations of the standard. *Courtesy National Fire Protection Agency*

Figure 2-8 NIOSH is the National Institute for Occupational Safety and Health, the federal agency that conducts research and makes recommendations for preventing workplace injuries. *Courtesy of the National Institute for Occupational Safety and Health*

provides information and field training. Many statistics about workplace injuries come from NIOSH research. Upon request, NIOSH will also evaluate the hazards of specific workplaces and recommend solutions.

Although NIOSH and OSHA share the common goal of worker safety, they are very different agencies. OSHA is involved in the creation and enforcement of safety rules and regulations, while NIOSH is involved in research and providing information. They work together to provide a safer work environment for US workers.

First Aid-CPR-AED

Administering First Aid means helping an individual who has been injured or is in need of help. It can include anything from helping to clean a cut and apply a band-aid to assisting a person who is suffering from heat stroke. CPR (cardiopulmonary resuscitation) is a lifesaving technique used on a person who has stopped breathing, or whose heart has stopped beating. An AED (Automated External Defibrillator) is a machine that uses electrical current to attempt to shock the heart into a normal rhythm. There are classes offered for First Aid, CPR, and AED individually, and there are also classes available that train a person in all three. Be sure the course you are planning to take is accredited. Upon completion of the training a certification card will be given. (Figure 2-9) The certification cards have an expiration date, so retaking a class is occasionally required.

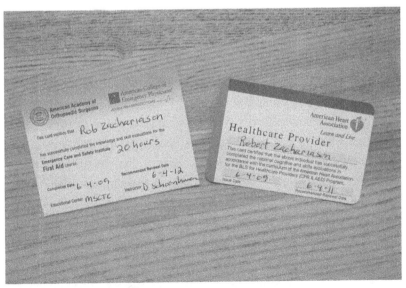

Figure 2-9 First Aid and CPR certification will expire after a couple of years, requiring a person to take a refresher class or retake the course. *Courtesy Delmar/Cengage Learning*

Emergencies requiring First Aid can happen at any time to anyone. The construction industry, however, has many hazards that people in other lines of work don't face, thus increasing the chance of someone becoming injured and requiring First Aid. Every person on a construction site should have current First Aid and CPR training, and many construction sites require that employees have current CPR, First Aid, and AED training. OSHA requires that at least one person on the job-site have current First Aid training.

You don't want to be in a situation where someone near you needs help and you don't know what to do; likewise, if you need help you would hope that someone around you has had the proper training. First Aid and CPR training is reasonably inexpensive, and can be completed in just a matter of hours. Many employers will offer training for their employees and cover the cost. If you are interested in taking the courses on your own initiative, the American Red Cross and other organizations offer inexpensive classes to the general public.

Summary

- The Occupational Safety and Health Administration (OSHA) is a federal organization whose purpose is to ensure the safety of people in the workplace.
- Employers must comply with all OSHA standards and regulations, and bear the responsibility of providing a workplace that is free from hazards that may cause serious injury or death to their employees. This includes having an effective safety program.
- Employees are responsible for following OSHA requirements as well as any other safety requirements the employer has implemented.
- The NFPA 70E (Standard for Electrical Safety in the Workplace) establishes safety standards for those installing or working on electrical systems.
- The National Institute for Occupational Safety and Health is an agency whose purpose is to help assure that working conditions for men and women are safe and healthful.

Review Questions

1. In what year did Congress pass the Occupational Safety and Health Act?
2. What is the purpose of OSHA?
3. What scenario is given OSHA's top inspection priority?
4. According to OSHA, if a person hasn't completed a task in _____, retraining is necessary.
5. What is the name of the NFPA 70E?
6. In what year was NFPA 70E first published?
7. What is the purpose of NFPA 70E?
8. What does the acronym NIOSH stand for?
9. What is an Automated External Defibrillator?

3

Electrical Shock

Objectives

- Describe how an electrical current affects the human body

- Describe the hazards associated with receiving an electrical shock

- Understand what determines the severity of an electrical shock

- Describe the steps necessary to remove electrical power

- Understand the procedures used to ensure equipment is in a de-energized state

- Describe the use and purpose of GFCI devices

▚ Introduction

Electrical shock is one of the most common workplace injuries. Electricians in particular are at a greater risk of electrical shock due to the nature of their work. Many electricians feel it is their job to work on live circuits, which couldn't be further from the truth. There is also a misconception that 120 volt circuits are harmless, but many people have lost their lives from coming into contact with a 120 volt circuit. OSHA prohibits working on any energized circuit of a voltage greater than 50 unless shutting off the power will create a greater hazard or is infeasible due to equipment design or operational limitations. In either case, if work is to be performed on a live circuit, the appropriate procedures must be followed and personal protective gear must be worn. This chapter will discuss the various dangers associated with receiving an electrical shock and how to safely work around electricity.

▚ Electricity and The Human Body

What is the difference between receiving an electrical shock and being electrocuted? Electrical shock happens when a negligible amount of current passes through the body. The effects of receiving a shock can range from being slight and hardly noticeable to very severe, causing serious injury or death. Electrocution is death resulting from an electrical shock. There are several possible consequences of receiving an electrical shock. Figure 3-1 lists a few of these dangers, all of which could end up being fatal.

Dangers associated with receiving an electric shock	
Ventricular Fibrillation	Heart cannot effectively pump blood
External Damage	Entrance and exit wounds will have burns
Internal Damage	Tissue damage to organs, nerves, muscles, etc.
Falls	Loose balance and fall from ladder, scaffold, etc.
Jerk Reaction	Pull hand away into sharp objects or moving machinery

Figure 3-1 This chart lists some of the dangers associated with receiving an electrical shock. *Courtesy Delmar/Cengage Learning*

What Determines The Severity of a Shock?

The severity of an electrical shock is determined by three factors: the amount of current, the path of the current, and the amount of time the current flows. Notice that the common element of all three factors is current. (Figure 3-2) At a current level of approximately 1 milliamp, a person will feel a slight tingling sensation. At a level of approximately 20 milliamps a person's muscles will contract, preventing them from letting go of the circuit. At a level of approximately 100 milliamps, the heart may fibrillate and stop beating. To put these current levels into perspective, it takes 500 milliamps to light a 120 volt 60 watt lamp—five times the amount of current necessary to cause ventricular fibrillation.

Amount of Current

The amount of current is the first determining factor in the severity of an electrical shock. As little as fifteen milliamps (.015A) of current can be lethal. The amount of current that flows through the body is

EFFECT OF ELECTRIC SHOCK		
	Current in milliamperes @ 60 hertz	
	Men	**Women**
• cannot be felt	0.4	0.3
• a little tingling—mild sensation	1.1	0.7
• shock—not painful—can still let go	1.8	1.2
• shock—painful—can still let go	9.0	6.0
• shock—painful—just about to point where you can't let go—called "threshold"—you may be thrown clear	16.0	10.5
• shock—painful—severe—can't let go—muscles immobilize—breathing stops	23.0	15.0

• ventricular fibrillation (usually fatal)
  length of time . . . 0.03 sec.
  length of time . . . 3.0 sec.
	1000
	100
		1000
		100

Figure 3-2 This chart lists the effects various current levels will have on the human body. *Courtesy Delmar/Cengage Learning*

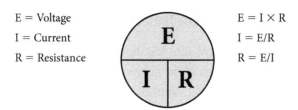

Figure 3-3 Ohm's law shows the relationship between voltage, current, and resistance. *Courtesy Delmar/Cengage Learning*

determined by the voltage of the power source, the resistance of the body, and the amount of available current.

Although voltage isn't considered a determining factor in the severity of a shock, it is one of the factors which determine how much current will flow. Ohm's law identifies the relationship between voltage, current, and resistance. (Figure 3-3) Voltage and current have a direct relationship: the higher the voltage, the higher the resulting current will be. For example, if a person with a resistance of 10,000 ohms were to receive a shock from a 120 volt power source, the current flow would be .012 amps. If the same person were to receive a shock from a 277 volt power source, the current flow would be .0227 amps. (See figure 3-4)

The resistance of the human body is another factor that determines how much current will flow. The resistance of a person's body can vary greatly; it depends on the part of the body that makes contact, whether a person has been perspiring, the amount of body fat, and many other factors. (Figure 3-5) The inside of the human body has a very low resistance: if a person was to have a conductor penetrate the

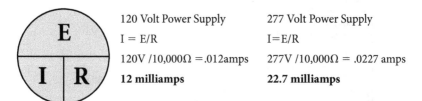

Figure 3-4 Voltage and current have a direct relationship. With a constant resistance, an increase in voltage will result in an increase in current. *Courtesy Delmar/Cengage Learning*

	Resistance (ohms)	
Condition	Dry	Wet
Finger touch	40,000 to 1,000,000	4,000 to 15,000
Hand holding wire	15,000 to 50,000	3,000 to 6,000
Finger-thumb grasp	10,000 to 30,000	2,000 to 5,000
Hand holding pliers	5,000 to 10,000	1,000 to 3,000
Palm touch	3,000 to 8,000	1,000 to 2,000
Hand around 1-1/2 inch pipe	1,000 to 3,000	500 to 1,500
Two hands around 1-1/2 inch pipe	500 to 1,500	250 to 750
Hand immersed		200 to 500
Foot immersed		100 to 300
Human body, internal, excluding skin	200 to 1,000	

This table contains data developed by Kouwenhoven and Milnor.

Figure 3-5 The resistance of a human body can vary. *Courtesy Delmar/Cengage Learning*

skin or touch a fresh wound, the current would have a fairly low resistance path. The skin is a person's last line of defense, and its resistance can vary greatly. A person with thick calluses will have a higher resistance than a person with baby-smooth hands. Likewise, a person with dry hands will have a higher resistance than a person with moist, sweaty hands. Regardless, there isn't enough resistance in a person's skin to protect against dangerous voltages.

Some power supplies will have a limited amount of current available. They may have extremely high voltages, but the current level will be limited to a value that isn't lethal. An example of this is a stun gun or TASER®. (Figure 3-6) They have voltages in the tens of thousands, but only a limited amount of available current. The shock administered will hurt and incapacitate a person while the shock is being received, but isn't lethal. Another example of this is static electricity. A shock from static electricity is typically between 4,000 and 6,000 volts, but has low current and therefore isn't lethal.

Figure 3-6 A stun gun used by correctional officers to incapacitate a person by administering an electrical shock. *Courtesy Delmar/Cengage Learning*

Current Path

The path of an electrical current is the second determining factor in the severity of a shock. An electrical current that passes through the chest cavity poses a greater threat of causing serious injuries to organs or death than if the current path is in and out of the same hand. This is not to say that it isn't dangerous to have a current flowing through a hand, but there is a much greater chance of lethal injury when the current passes through vital organs.

Current that flows near the heart may cause the heart to fibrillate. The amount of current flow which can cause the heart to fibrillate is generally in the range of 100–200 milliamps. When a heart fibrillates it begins to quiver and is unable to pump blood. The current paths that are most likely to cross the heart are hand-to-hand and hand-to-foot contact. (Figure 3-7)

A person whose heart is fibrillating from an electrical shock will lose consciousness and require immediate medical attention. A heart that is fibrillating will typically not be able to recover its normal rhythm on its own, resulting in death. An Automated External

Defibrillator (AED) will be needed to shock the heart back into a normal rhythm. (Figure 3-8)

Length of Time

The length of time a person is receiving a shock is the third factor which determines the severity of an electrical shock. The longer a person has current flowing through their body, the more damage will occur. At lower current levels, the natural tendency is to jerk away

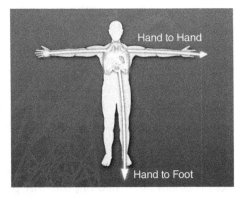

Figure 3-7 Current which crosses vital organs can cause serious internal injuries or death. *Courtesy Delmar/Cengage Learning*

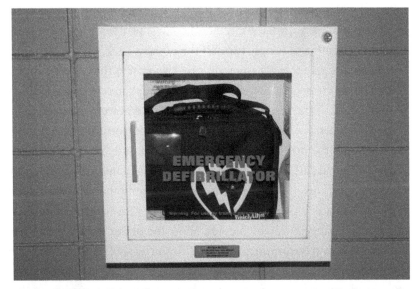

Figure 3-8 Some businesses will have Automated External Defibrillators in various locations on the premises. *Courtesy Delmar/Cengage Learning*

when receiving a shock, limiting the shock time. At approximately 20 milliamps, however, a person loses the ability to pull away due to muscular paralysis, and the current will continue to flow through the person, burning a path through the body. A person can receive serious internal burns and tissue damage from this current flow.

Witnessing a Shock

What should you do if you witness someone receiving a shock? Hopefully the person is able to release themselves from the circuit. If the person is stuck on the circuit and not able to let go, the first step for the witness is to assess the situation. What is the voltage that the person is in contact with? If the person is in contact with a downed power line or medium voltage line, stand clear. Any attempt you make to physically remove the person will undoubtedly cause you to be in contact with the circuit as well—your safety must come first. In the situation of a downed power line or medium voltage line, unless there is an accessible disconnect that can be turned off, the only thing you can do is call for help. If it is determined that the person is in contact with a typical residential or commercial voltage (which is considered low voltage by National Electric Code standards), the first thing is to try to remove the person from the circuit. The safest way to remove a person from the circuit is to simply turn off the power.

There are many ways a person can disconnect the power supply. It may be shutting off a switch, disconnect, breaker, or simply unplugging the cord. If the means of disconnecting the power isn't close then it may be necessary to physically remove the person from the circuit. NEVER grab the person to pull them off the circuit. You may become a part of the circuit as well and become incapacitated. If you need to push them off the circuit be sure to use something that is nonconductive, such as a dry piece of wood. Remember that most new lumber still has moisture in it, which could allow the board to be conductive and result in you becoming part of the circuit as well. Always put your own safety first. You won't be able to help at all if you also become stuck in the circuit. Once a person has been safely removed from the circuit, administer First Aid.

In the event of an electrical shock, always seek medical attention. Whether the person was able to free themselves or had to be removed from the circuit by others, they need to go and have the injury checked out by a medical professional. There may be internal damage that can't be seen or felt, and some of the effects may not be immediate. Ventricular fibrillation can occur hours or even days after the shock. It is also good to have medical documentation in case of any future problems a person may have that could be related to the incident.

How to Work Safely Around Electricity

If a person is following the appropriate safety guidelines, there should be no possibility of an electrical shock. OSHA and the NFPA 70E are devoted to worker safety and have provided all the necessary requirements to ensure that a worker goes home safely at the end of the day.

OSHA and the NFPA 70E are very clear on the fact that there are very few situations where it is permitted to work on energized equipment. The only time that energized work is permitted is when disconnecting power creates a greater hazard, or if it is infeasible due to equipment design and operational limitations.

The most important thing, that can't be stressed enough, is <u>DO NOT WORK ON ENERGIZED CIRCUITS!</u> If all potential power supplies have been removed or discharged, there is no chance of getting an electrical shock. There are several important steps to follow to ensure that there is no potential voltage.

1. Planning;
2. Disconnect the power;
3. Use the appropriate lockout-tagout procedures to ensure the power isn't turned back on;
4. Release any stored energy;
5. Verify that there is no potential voltage, using the proper procedures (PPE, meter requirements);
6. Ground conductors and equipment.

Planning

Every time a piece of equipment is going to be de-energized for service there is planning involved. For smaller jobs this may be as simple as figuring out where the switch is located, while others may require the use of schematics, wiring diagrams, or blueprints. It is also necessary to communicate with workers who may be affected by the power being removed.

Disconnect the Power

Disconnecting the power sounds easy enough, but there are several ways to accomplish this and a few scenarios to keep in mind. This chapter is on electrical shock so that is the focus, but don't forget that there may be other forms of energy that may have to be locked out, such as hydraulics, air, gas, steam, etc.

Switch

When working on equipment such as lights, disconnecting the power may be a simple as shutting off a switch. What one has to keep in mind is that although the switch may remove power to the light, there may still be potential voltage in the box above the light. Another issue is the possibility of someone walking into the room and turning on the switch before they notice you are working on the light. There are lockout devices which are designed to prevent a toggle switch from being turned on. (Figure 3-9) Some areas have multiple switches feed a light: if this is the case, care must be taken to ensure that all of these switches are secure against being turned on. If there is any concern that a switch may be turned on, or that there may be power in the box even with the switch in the "off" position, it is better to be safe and disconnect the circuit at the electrical panel.

Breaker

Many times the entire circuit needs to be shut down. This is done by shutting off the circuit breaker or removing a fuse. It is not safe to shut off a breaker or remove a fuse while it is under load. Contacts being

Figure 3-9 A lockout device to prevent a switch from being turned on.
Courtesy Delmar/Cengage Learning

opened with high current levels will cause excessive arcing which could damage the equipment, shorten its life, or even lead to a dangerous arcing incident. To eliminate this danger, the equipment that is fed from the circuit being shut down must be turned off before the breaker is operated or the fuse is removed. The electrical panel is often in a different room than the work being done, so it is imperative that the appropriate lockout-tagout procedures be followed to ensure that the circuit doesn't get turned back on. (Figure 3-10)

Disconnect

Most equipment will have a means of disconnection within sight of it. As with breakers and fuses, it is not safe to shut off a disconnect while it is under load. Contacts being opened with high current levels will cause excessive arcing which could damage the equipment, shorten its life, or lead to a dangerous arcing incident. To eliminate this danger, the equipment fed from the disconnect being shut down must be turned off before the disconnect is operated. After shutting off the disconnect switch, be sure to use the appropriate lockout-tagout procedures to ensure that it doesn't get turned back on. (Figure 3-11)

Figure 3-10 A lockout device to prevent a circuit breaker from being turned on. *Courtesy Delmar/Cengage Learning*

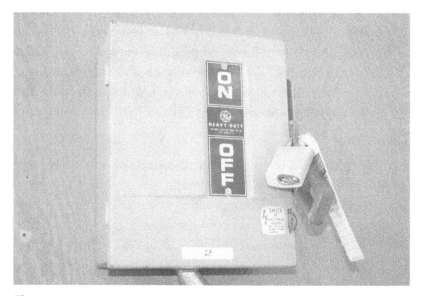

Figure 3-11 A lockout device to prevent a disconnect switch from being turned on. *Courtesy Delmar/Cengage Learning*

Cord and Plug

Some equipment is simply plugged into a receptacle. In this case all that is necessary to disconnect the equipment is to unplug it. Remember to shut the equipment off before removing the plug to prevent excessive arcing that could produce a dangerous situation or damage the equipment. There are caps that can be locked over the end of the cord to ensure that it doesn't get plugged back in while you are working on it. (Figure 3-12)

Lockout-Tagout

Locking and tagging out equipment is as important as shutting it off; this procedure prevents someone from turning the equipment back on. Specific lockout-tagout procedures will be included in a company's safety plan. Locking out equipment provides a physical restraint that prevents the power from being turned back on, and tagging out equipment provides information on who is performing the work, the date and time, reason for work, etc.

Figure 3-12 A lockout device to prevent a cord from being plugged in. *Courtesy Delmar/Cengage Learning*

Stored Energy

Some equipment may have electrical, kinetic, or potential energy that must be released or removed. Electrical energy stored may be in batteries that must be disconnected or capacitors that must be discharged. Be sure to follow the correct procedures for disconnecting batteries and discharging capacitors, which will include verifying that the power has been disconnected before discharging capacitors, and wearing the appropriate personal protective equipment, as the components may have a potential voltage. This chapter is on electrical shock so that is the focus, but don't forget that equipment may have kinetic energy (spinning wheels, blades, etc.) or potential energy (springs, elevated weight, etc.) that must also be addressed.

Verify There Is No Potential Voltage

Before considering equipment de-energized, the absence of potential voltage must be verified with an electrical meter. (Figure 3-13) There are several reasons why there could be potential voltage even though you have just disconnected the power.

- It is possible that the wrong breaker or disconnect was shut off;
- The power was turned back on;
- An induced voltage;
- A voltage backfed from other circuits or equipment;
- Stored energy (batteries or capacitors).

While performing this test, a person must wear the appropriate personal protective equipment and assume that the power is still on. Always verify that the meter is working properly by checking a known voltage before using it to check for the absence of voltage on the supposedly de-energized equipment; a faulty meter or meter lead could prove fatal. While checking for voltage, always check from phase to phase and from phase to ground on all conductors.

Ground Conductors

If there is potential for an induced voltage or for a wire to become energized while working on the circuit, then all phase conductors and

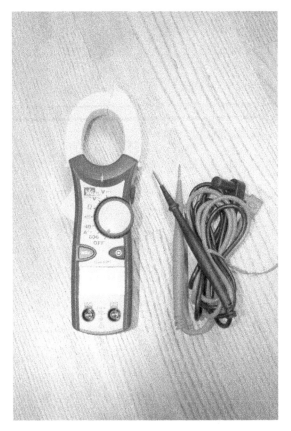

Figure 3-13 A voltmeter can be used to verify the absence of voltage. *Courtesy Delmar/Cengage Learning*

equipment must be grounded before work commences. Be sure the conductor used to ground the conductors and equipment is sized to handle the potential fault. (Figure 3-14)

Ground Fault Circuit Interrupter (GFCI)

Electrical shock due to a faulty tool or cord is one of the leading causes of electrical shock on a construction site. One way to prevent electrocution while using electrical equipment is by using Ground Fault Circuit Interrupters (GFCI). Ground Fault Circuit Interrupters will disconnect the circuit in the event of a ground fault before the current level and length of time the current flows become high enough to be lethal.

Figure 3-14 Ground set. *Courtesy of salisbury by Honeywell*

OSHA and the National Electrical Code (NEC) require all 125 volt 15, 20, and 30 amp receptacles used for construction, maintenance, remodeling, repair, demolition, or similar activities to be GFCI-protected.

GFCI devices monitor the current going out and coming back in the circuit. If there is current missing from the circuit it is possible that it is going through a human body to ground. (Figure 3-15) An imbalance of four to six milliamps will cause a Class A GFCI to disconnect the power to the circuit, and it will disconnect the circuit in as little as 1/40th of a second. (Figure 3-16) A person may still receive a shock, but the circuit will be opened before the shock becomes lethal. It is important to realize that a GFCI will not trip with a line to neutral fault. If a person is receiving a shock from contacting a

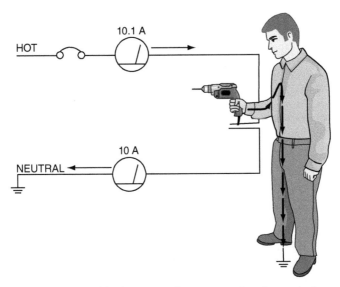

Figure 3-15 A ground fault occurs when current flow leaves its intended path and flows to ground. *Courtesy Delmar/Cengage Learning*

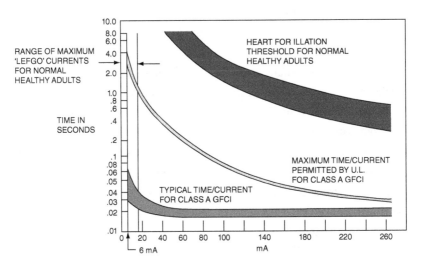

• CLASS 'A' GFCI :
• TRIPS WHEN CURRENT TO GROUND IS 6 mA OR GREATER.
• DOES NOT TRIP WHEN CURRENT TO GROUND IS LESS THAN 4 mA.
• MAY OR MAY NOT TRIP WHEN THE CURRENT TO GROUND IS BETWEEN 4 mA AND 6 mA.

Figure 3-16 This illustration is a time/current curve which shows the tripping characteristics of a typical Class A GFCI. *Courtesy Delmar/Cengage Learning*

Figure 3-17 GFCI circuit breaker. *Courtesy Delmar/Cengage Learning*

hot and neutral conductor, there is no current leaking to ground, and therefore no imbalance detected by the GFCI.

The three main types of GFCI devices used are: GFCI circuit breaker, GFCI receptacle, and GFCI cords or splitters. A GFCI circuit breaker is installed in the electrical panel and offers protection for the entire circuit that it feeds. (Figure 3-17) A GFCI receptacle, which is the most commonly used GFCI device, is installed in place of a regular receptacle. GFCI receptacles can not only protect the equipment plugged into them, but can also protect receptacles that are fed downstream. (Figure 3-18) In the event that the circuit a person is going to use doesn't have GFCI protection, a cord that has built-in GFCI protection can be used. (Figure 3-19)

Ground Fault Circuit Interrupters have a test button built into the device to verify that it is working properly. They should always be tested according the manufacturer's instructions, which is typically monthly for receptacles and breakers, and prior to each use in the case of cords with built-in GFCI protection.

Figure 3-18 GFCI receptacle outlet. *Courtesy Delmar/Cengage Learning*

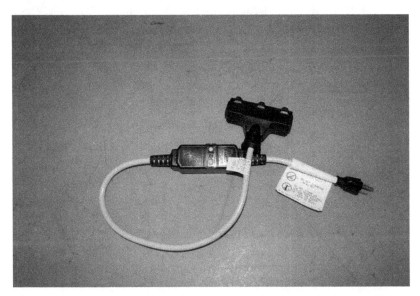

Figure 3-19 Two types of portable plug-in cord sets with built-in GFCI protection. *Courtesy Delmar/Cengage Learning*

Summary

- OSHA and the NFPA prohibit working on energized circuits and equipment.
- Electrical shock may result in electrocution, external burns, internal burns, tissue damage, falls, or moving hand or body into moving or sharp objects.
- The severity of an electrical shock is determined by the amount of current, the path of the current, and the amount of time the current flows.
- If a person is receiving a shock, they must be safely removed from the circuit, and the appropriate First Aid procedures followed.
- There are several steps that must be followed when disconnecting circuits or equipment for servicing:
  - Carefully plan how the job is to be performed and communicate to any workers that may be affected by the power being disconnected;
  - Shut down equipment before removing the power;
  - Use the appropriate lockout-tagout procedures;
  - Release stored energy;
  - Verify absence of voltage;
  - Use safety grounds when necessary.
- GFCI devices will disconnect the circuit in the event of a ground fault before the current levels become lethal.
- The three types of GFCI devices typically used on construction sites are: GFCI breakers, GFCI receptacles, and cords or splitters with built-in GFCI protection.

Review Questions

1. What is the difference between a shock and electrocution?
2. What are the three factors which determine how severe an electrical shock is?
3. List four hazards associated with receiving an electrical shock.

4. What effect does voltage have on the amount of current passing through the body?
5. What are the six steps involved with safely removing the power?
6. At _____ of electrical current a human heart will fibrillate.
7. OSHA prohibits working on energized circuits of _____ or more.
8. If a person receives a shock they should always seek _____.
9. Class A GFCI devices will open a circuit in the event of a ground fault in the current range of _____.

4

Arcing Incidents

Objectives

- Describe the dangers associated with an arcing incident

- Understand the impact that an arcing incident will have on persons and property

- Describe situations that could result in an arcing incident

- List the steps necessary to protect a person from an arcing incident

🔳 Introduction

When most people think of the dangers associated with electricity, their first thought is electrical shock. Electrical shock is definitely a danger to be avoided, but it isn't the only hazard associated with electricity. Arcing incidents are another hazard of working around electricity, one that most people are unaware of. They can be extremely destructive, causing major property damage, severe bodily injuries, or death. This chapter aims to explain the dangers of arcing incidents and how to protect yourself and fellow workers from them.

🔳 Arcing Incident

An arcing incident will begin with an arcing fault and will escalate to an arc flash and arc blast. Arcing incidents can vary considerably; from the very minor, causing no property damage or bodily injury, to the devastating, causing major property damage and severe injuries or death. (Figure 4-2)

Arcing Fault

An arcing fault is the flow of electricity through ionized air between two conductive objects with different electrical potential. The arcing fault may be due to two conductors getting too close or touching; a conductive foreign object crossing two conductors; cracks or imperfections in the insulation; current flowing across a film or conductive dust; the melting or burning away of a conductor due to a loose connection or high current; or faulty equipment.

Figure 4-1 Arch flash hazard symbol. *Courtesy Delmar/Cengage Learning*

Figure 4-2 Picture of arcing incident. *Courtesy of Salisbury by Honeywell*

There are several factors which determine whether an arcing fault turns into a serious situation involving an arc flash and arc blast:

- How much fault current is available: the generator or utility transformer will have a maximum amount of current available in the case of a short circuit or ground fault. (Figure 4-3) This is the available fault current. The higher the available fault current, the more potential energy present in the case of an arcing incident. The amount of fault current available at a customer-owned generator can be found by contacting the manufacturer, while the fault current at the utility transformer can be ascertained by contacting the utility company. The available fault current is a complicated calculation which is typically performed by an electrical engineer.

Figure 4-3 Pad-mounted utility transformer. *Courtesy Delmar/Cengage Learning*

- Distance to fault from power source: how far the fault is from the generator or utility transformer. Utility transformers have a maximum amount of fault current they can deliver, and every length of conductor the current has to pass through will limit it to some degree. The further from the utility transformer, the less fault current available. Switchgear that is very close to the utility transformer and has very large conductors or bussing connecting the two will have a very high level of fault current available.
- Overcurrent protective devices: whether the overcurrent protection is sized properly and has the correct short circuit current ratings. Properly-sized overcurrent devices (breakers and fuses) will typically be able to handle the fault current passing through them and will safely open the circuit. (Figure 4-4) Short circuit current calculations are performed by electrical engineers and are beyond the scope of this text. However, electricians must be sure to install equipment with the proper ratings.

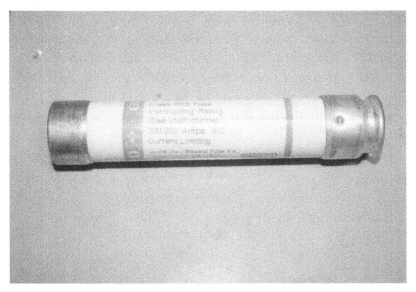

Figure 4-4 Interrupting rating on cartridge fuse. *Courtesy Delmar/Cengage Learning*

- Location of fault in the circuit: if there is a load in series with the fault it will limit the current flow. It may still arc and possibly cause property damage or a fire, but it will more than likely not cause a dangerous explosion.

Arc Flash

An arc fault with enough potential energy will create an arc flash, which is a rapid release of energy due to an arcing fault. An arc flash superheats the air that is conducting current and turns it into plasma. This plasma has a temperature which is up to 35,000 degrees Fahrenheit (four times hotter than the surface of the sun) and superheats the surrounding air, causing it to expand at a very high rate. At the high temperatures found in an arc flash, metal is instantly vaporized. Copper, when vaporized, expands to 67,000 times its volume, creating an immense shock wave. Due to the expanding shock wave from superheated air and vaporized metals, any metal or other shrapnel—whether molten or solid—will be propelled at very high velocities.

The amount of heat created in an arc flash will severely burn a human body. Arc flash survivors who lacked appropriate personal protective equipment (PPE) can sustain second- and third-degree burns over major portions of their body. They can end up in a burn unit for months, and may never be the same. Each year, more than 2,000 people are admitted into burn centers as a result of arcing incidents.

The arc flash will also create an intense light which can be damaging to the eyes—it may be compared to staring at the sun or watching a welder. It has been shown that exposure to the intense light from an arc flash can lead to future eye problems, such as cataracts.

Arc Blast

An arc blast is the rapid expansion of air and metal associated with an arc flash. Under the right conditions, an arc blast can occur with the same power as dynamite. The pressure wave associated with an arc blast is capable of rupturing eardrums, and even crushing lungs or causing other fatal injuries. The blast is capable of throwing a person into energized components, or across a room. Since many electrical rooms are small, this often results in a person being slammed into a wall.

Many times an arc blast occurs in an enclosure, such as a panelboard or switchboard, which has sides and a back. (Figure 4-5) The blast may blow the enclosure apart and create dangerous flying debris, or the enclosure may project the blast forward, intensifying the effects. The molten metal and other shrapnel are projected at speeds of up to 700 miles per hour and will penetrate the human body.

The sound associated with a minor arcing fault is loud; it sounds very similar to a shotgun being fired and will leave your ears ringing. The sound from a major arc blast, on the other hand, is deafening. It has been recorded at over 160 decibels, enough to cause permanent hearing damage.

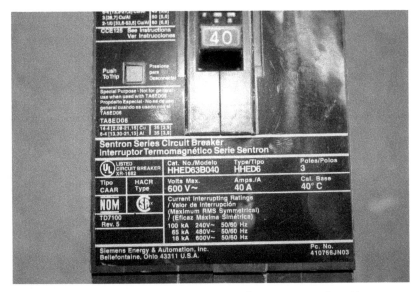

Figure 4-5 Interrupting rating on circuit breaker. *Courtesy Delmar/Cengage Learning*

Causes of Arcing Incidents

There are several reasons why an arcing incident can occur. Unfortunately, they often result from human error, but may also be due to a piece of equipment that is faulty or simply fails.

A few examples of human errors which may cause an arcing incident:

- A screwdriver or other tool slips while work is being done in an energized electrical panel and shorts together two conductive parts with a difference in potential. (Figure 4-6)
- A metal object falls into a panel and shorts together two conductive parts with a difference in potential. (Figure 4-7)
- The overcurrent protection may have been improperly sized, having a short circuit rating lower than the available short circuit current.
- A fuse or breaker may have been replaced taking into account only the ampacity of the breaker, not the short circuit rating.

Figure 4-6 Screwdriver creating a short circuit in a meter socket. *Courtesy Delmar/Cengage Learning*

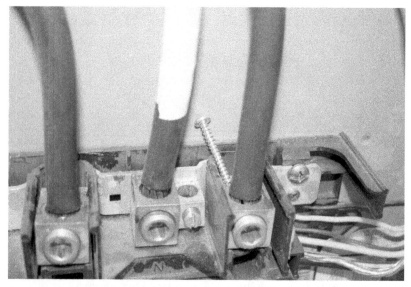

Figure 4-7 Screw creating a short circuit in a panelboard. *Courtesy Delmar/ Cengage Learning*

A few examples of situations not directly tied to human error where arcing incidents can occur.

- Overcurrent device failure:
  - May be triggered by the vibrations from simply opening the cover on the panelboard, or by turning the overcurrent device on or off.
  - May be caused by internal device failure
- Breaks or cuts in a conductor's insulation which contact the side of an enclosure, fitting, or raceway.
- Corrosion or a film of dust, oil, or other contaminants in equipment which conduct current and leads to an arcing incident.

Arc faults will occur without warning and happen so fast that there is no way to get clear or to respond. The only defense is to be well-prepared. It is our job to be thinking about what could occur so that we take the necessary steps to safeguard ourselves for the worst-case scenario. This means understanding the hazard involved, taking the appropriate steps, and wearing the necessary PPE.

Arcing Incident Protection

As electricians we must safeguard ourselves when working around equipment that has an arc flash potential. The first step is to determine what the potential energy could be in the case of a fault so the appropriate PPE can be worn as a precaution. This can be done by having an arc flash hazard analysis performed, usually by an electrical engineer. The National Electrical Code requires that all non-dwelling equipment that may require servicing with the power on carry an arc flash label if there is a potential hazard. (Figure 4-8) Examples of equipment that would require this label include switchboards, panelboards, motor control centers, industrial control centers, meter sockets, etc. The NEC only requires that the label identifies the existence of a hazard, and does not require the label to indicate the hazard/risk category or indicate what level of PPE is necessary. The NFPA 70E, however, does address this issue.

Figure 4-8 Arc flash hazard warning label. *Courtesy Delmar/ Cengage Learning*

The NFPA 70E requires that equipment either be labeled in the field with the available incident energy or the required level of PPE necessary. Labels installed on more modern equipment will indicate not only the incident energy level and required PPE, but also describe the shock hazard and boundary distances. (Figure 4-9) Electricians must keep in mind that this wasn't always a requirement, so equipment in older buildings may not have this label even though there may be a hazard.

Arc Flash Protection Boundary

The NFPA 70E bases the requirements for wearing PPE, which protects workers from arcing incidents, on the arc flash protection boundary. This boundary is the extent of the area around equipment within which a person without appropriate PPE would receive a second-degree burn in the event of an arc flash. The arc flash protection boundary is found by performing a calculation, which is typically done by an electrical engineer. The bare minimum boundary on equipment having an arc fault hazard is four feet. The boundary is a sphere which surrounds equipment having the potential to produce an arc flash or blast. (Figure 4-10) Any work that is to be performed within this boundary will require the use of PPE. Keep in mind that the minimum four-foot arc flash boundary applies to equipment with a low arc flash hazard; equipment with a greater hazard will have much more extensive boundaries.

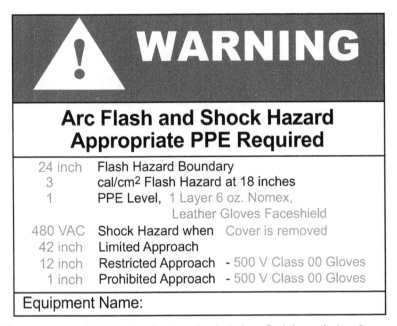

⚠ **WARNING**

Arc Flash and Shock Hazard
Appropriate PPE Required

24 inch	Flash Hazard Boundary
3	cal/cm² Flash Hazard at 18 inches
1	PPE Level, 1 Layer 6 oz. Nomex, Leather Gloves Faceshield
480 VAC	Shock Hazard when Cover is removed
42 inch	Limited Approach
12 inch	Restricted Approach - 500 V Class 00 Gloves
1 inch	Prohibited Approach - 500 V Class 00 Gloves

Equipment Name:

Figure 4-9 Arc flash label indicating shock and arc flash boundaries. *Courtesy Delmar/Cengage Learning*

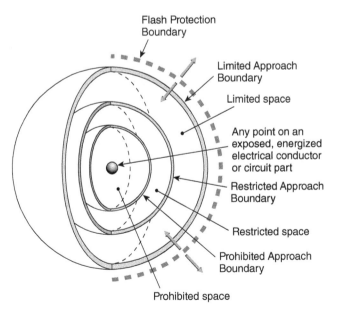

Figure 4-10 Shock and arc flash protection boundaries. *Courtesy Delmar/Cengage Learning*

There are two ways to determine the necessary PPE required for working inside the arc flash protection boundary. The first is to perform an arc flash hazard calculation, which will indicate the amount of potential energy and dictate the required PPE. The completed calculation will identify the amount of potential energy and therefore indicate the size of the arc flash protection boundary as well as the necessary PPE. This information will often be posted on the arc flash hazard label.

The second method is to use the NFPA 70E tables to determine the hazard/risk level and the appropriate PPE needed.

NFPA 70E Tables

The necessary level of PPE can be found in Table 130.7(C)(9) of the NFPA 70E. (Figure 4-11) This table has a list of various tasks and the voltage and type of equipment that may be worked on. Each task is given a hazard/risk category based on the nature of the work being performed. It will also indicate if rubber insulating gloves and insulated hand tools are required. The hazard/risk is divided into categories from 0-4, with 0 being the lowest hazard with the minimum amount of PPE required, and 4 being the most dangerous, with higher levels of PPE required.

The next step is to determine what PPE is necessary for each specific category. NFPA 70E Table 130.7(C)(10) lists the hazard/risk categories and the PPE necessary for each. (Figure 4-12)

Note that turning on a breaker in a 240v panel with the covers on is a hazard/risk category 0 and requires the wearing of non-melting long-sleeve shirt and pants, safety glasses, hearing protection, and leather gloves. Simply turning on a breaker can be hazardous! If the circuit fed from the breaker has a fault and the overcurrent device isn't sized properly or is faulty, it could potentially cause an arcing incident. It is extremely important to wear the appropriate PPE, stand off to the side, and turn your head away when operating breakers, disconnects, and the like.

Table 130.7 (C)(9) Hazard/Risk Category Classifications and Use of Rubber Insulating Gloves and Insulated and Insulating Hand Tools

Tasks Performed on Energized Equipment	Hazard/Risk Category	Rubber Insulating Gloves	Insulated and Insulating Hand Tools
Panelboards or Other Equipment Rated 240 V and Below — Note 1			
Perform infrared thermography and other non-contact inspections outside the restricted approach boundary	0	N	N
Circuit breaker (CB) or fused switch operation with covers on	0	N	N
CB or fused switch operation with covers off	0	N	N
Work on energized electrical conductors and circuit parts, including voltage testing	1	Y	Y
Remove/install CBs or fused switches	1	Y	Y
Removal of bolted covers (to expose bare, energized electrical conductors and circuit parts)	1	N	N
Opening hinged covers (to expose bare, energized electrical conductors and circuit parts)	0	N	N
Work on energized electrical conductors and circuit parts of utilization equipment fed directly by a branch circuit of the panelboard	1	Y	Y
Panelboards or Switchboards Rated >240 V and up to 600 V (with molded case or insulated case circuit breakers) — Note 1			
Perform infrared thermography and other non-contact inspections outside the restricted approach boundary	1	N	N
CB or fused switch operation with covers on	0	N	N

Figure 4-11 Table 130.7(C)(9), 2009 NFPA 70E. Reprinted with permission from NFPA 70™, *National Electrical Code®*, Copyright© 2007, National Fire Protection Association, Quincy, MA 02269

Table 130.7(C)(10) Protective Clothing and Personal Protective Equipment (PPE)

Hazard/Risk Category	Protective Clothing and PPE
Hazard/Risk Category 0 Protective Clothing, Nonmelting (according to ASTM F 1506-00) or Untreated Natural Fiber	Shirt (long sleeve) Pants (long)
FR Protective Equipment	Safety glasses or safety goggles (SR) Hearing protection (ear canal inserts) Leather gloves (AN) (Note 2)
Hazard/Risk Category 1 FR Clothing. Minimum Arc Rating of 4 (Note 1)	Arc-rated long-sleeve shirt (Note 3) Arc-rated pants (Note 3) Arc-rated coverall (Note 4) Arc-rated face shield or arc flash suit hood (Note 7) Arc-rated jacket, parka, or rainwear (AN)
FR Protective Equipment	Hard hat Safety glasses or safety goggles (SR) Hearing protection (ear canal inserts) Leather gloves (Note 2) Leather work shoes (AN)
Hazard/Risk Category 2 FR Clothing, Minimum Arc Rating of 8 (Note 1)	Arc-rated long-sleeve shirt (Note 5) Arc-rated pants (Note 5) Arc-rated coverall (Note 6) Arc-rated face shield or arc flash suit hood (Note 7) Arc rated jacket, parka, or rainwear (AN)
FR Protective Equipment	Hard hat Safety glasses or safety goggles (SR) Hearing protection (ear canal inserts) Leather gloves (Note 2) Leather work shoes
Hazard/Risk Category 2* FR Clothing, Minimum Arc Rating of 8 (Note 1)	Arc-rated long-sleeve shirt (Note 5) Arc-rated pants (Note 5) Arc-rated coverall (Note 6) Arc-rated arc flash suit hood (Note 10) Arc-rated jacket, parka, or rainwear (AN)
FR Protective Equipment	Hard hat Safety glasses or safety goggles (SR) Hearing protection (ear canal inserts) Leather gloves (Note 2) Leather work shoes

Figure 4-12 Table 130.7(C)(10), 2009 NFPA 70E. Reprinted with permission from NFPA 70™, *National Electrical Code*®, Copyright© 2007, National Fire Protection Association, Quincy, MA 02269

Hazard/Risk Category	Protective Clothing and PPE
Hazard/Risk Category 3 **FR** Clothing, Minimum Arc Rating of 25 (Note 1)	Arc-rated long-sleeve shirt (AR) (Note 8) Arc-rated pants (AR) (Note 8) Arc-rated coverall (AR) (Note 8) Arc-rated arc flash suit jacket (AR) (Note 8) Arc-rated arc flash suit pants (AR) (Note 8) Arc-rated arc flash suit hood (Note 8) Arc-rated jacket, parka, or rainwear (AN)
FR Protective Equipment	Hard hat FR hard hat liner (AR) Safety glasses or safety goggles (SR) Hearing protection (ear canal inserts) Arc-rated gloves (Note 2) Leather work shoes
Hazard/Risk Category 4 FR Clothing, Minimum Arc Rating of 40 (Note 1)	Arc-rated long-sleeve shirt (AR) (Note 9) Arc-rated pants (AR) (Note 9) Arc-rated coverall (AR) (Note 9) Arc-rated arc flash suit jacket (AR) (Note 9) Arc-rated arc flash suit pants (AR) (Note 9) Arc-rated arc flash suit hood (Note 9) Arc-rated jacket, parka, or rainwear (AN)
FR Protective Equipment	Hard hat FR hard hat liner (AR) Safety glasses or safety goggles (SR) Hearing protection (ear canal inserts) Arc-rated gloves (Note 2) Leather work shoes

AN = As needed (optional)

AR = As required

SR = Selection required

Notes:

1. See Table 130.7(C)(11). Arc rating for a garment or system of garments is expressed in cal/cm².

2. If rubber insulating gloves with leather protectors are required by Table 130.7(C)(9), additional leather or arc-rated gloves are not required. The combination of rubber insulating gloves with leather protectors satisfies the arc flash protection requirement.

3. The FR shirt and pants used for Hazard/ Risk Category 1 shall have a minimum arc rating of 4.

4. Alternate is to use FR coveralls (minimum arc rating of 4) instead of FR shirt and FR pants.

5. FR shirt and FR pants used for Hazard/ Risk Category 2 shall have a minimum arc rating of 8.

6. Alternate is to use FR coveralls (minimum arc rating of 8) instead of FR shirt and FR pants.

7. A face shield with a minimum arc rating of 4 for Hazard/Risk Category 1 or a minimum arc rating of 8 for Hazard/Risk Category 2, with wrap-around guarding to protect not only the face, but also the forehead. ears, and neck (or, alternatively, an arc-rated arc flash suit hood), is required.

8. An alternate is to use a total FR clothing system and hood, which shall have a minimum arc rating of 25 for Hazard/Risk Category 3.

9. The total clothing system consisting of FR shirt and pants and/or FR coveralls and/or arc flash coat and pants and hood shall have a minimum arc rating of 40 for Hazard/Risk Category 4.

10. Alternate is to use a face shield with a minimum arc rating of 8 and a balaclava (sock hood) with a minimum arc rating of 8 and which covers the face, head and neck except for the eye and nose areas.

Figure 4-12 Continued.

Personal Protective Equipment

Personal protective equipment is the final level of protection when it comes to an arc flash. A label on equipment that has had an arc flash hazard analysis may indicate the arc flash hazard as well as the necessary PPE. The NFPA 70E has tables which indicate the appropriate level of PPE based on the job being performed and the arc flash hazard. Be sure to read the label or check the NFPA tables to ensure that the correct PPE is being worn.

Chapter 5 discusses working on energized electrical equipment and covers the steps and To be verified necessary to work within the arc flash boundary.

Summary

- Arcing incidents are devastating events which can produce property damage and personal injury or even death.
- An arc flash creates temperatures greater than 35,000 degrees and intense light that can cause damage to the eyes.
- The sound associated with an arcing incident can be 160 decibels, enough to cause permanent hearing damage.
- Arc blasts project molten metal and shrapnel at speeds that can penetrate the human body.
- An arc flash hazard analysis will determine the amount of potential energy, which will in turn determine the appropriate PPE.

Review Questions

1. List three scenarios that could result in an arc flash.
2. What is the temperature of an arch flash?
3. Vaporized copper expands to _____.
4. List four damaging effects of an arc flash.

5. How can the arc flash potential be found?
6. Describe the arc flash protection boundary.
7. Without having an arc flash hazard analysis, the minimum arc flash protection boundary is _____.
8. _____ _____ is the last line of defense against an arcing incident.

CHAPTER 5

Working on Energized Equipment

Objectives

- Describe situations that may necessitate working on live circuits or equipment

- Describe the various approach boundaries

- Understand how the NFPA 70E tables apply to the various approach boundaries

- Describe an energized work permit

- Describe the personal protective equipment used to prevent electrical shock

- Describe the personal protective equipment used to prevent injuries from an arcing incident

- Describe the proper care for the various types of personal protective equipment

Introduction

There are very few scenarios that justify working on equipment that is energized, and every effort must be taken to disconnect the power before work begins. Electrical shock, burns, and death are just a few of the real dangers posed by energized equipment. There are, however, a few scenarios where the circuit must be energized when the work is being performed. This chapter is intended to describe the steps that must be taken to ensure a worker's safety while working on energized equipment.

A company's safety plan should detail the proper procedures and personal protective equipment (PPE) necessary to safely complete work on energized equipment.

When is it Permitted to Work on Energized Equipment?

OSHA and the NFPA 70E state that work should never be performed on energized electrical equipment unless disconnecting the power will create additional or increased hazards, or if it is infeasible due to equipment design or operational limitations. It is also very clear that only qualified persons are permitted to work on or near exposed energized electrical equipment. According to Article 100 of the NFPA 70E, a qualified person is: "one who has the skills and knowledge related to the construction and operation of the electrical equipment and installations and has received safety training to recognize and avoid the hazards involved."

Examples of situations where disconnecting the power may create an additional or increased hazard include:

- Life support equipment;
- Emergency alarm systems;
- Ventilation systems in hazardous locations.

Examples of situations where it may be infeasible to disconnect the power include:

- Diagnosing and testing;
- Equipment that is part of a continuous process, where the entire process would have to be shut down.

The only situation listed above where the circuit must be energized is diagnosing and testing. In order to take an accurate measurement of voltage and current, the circuit must be energized. This does not mean that the power must be on every time a person is diagnosing and testing. Some scenarios allow the circuit to be shut off, the test meter inserted into the circuit, and then the circuit turned back on to take the measurement. This may not always be practical, but it is the safest way to take measurements.

With the exception of diagnosing and testing, most situations can accommodate the power being turned off, given enough notice. After all, if an accident were to happen while work was being carried out on a piece of energized equipment, the power would be shut off regardless, possibly for an extended period of time.

All this means that there are very few situations that warrant leaving equipment energized while it is being worked on. Shut the power off! It may save your life.

Approach Boundaries

Before discussing the steps that must be taken to ensure a person's safety while working around energized electrical equipment, the approach boundaries listed in the NFPA 70E must be discussed. A person who is going to perform work on or near energized equipment must understand all the approach boundaries. There are several boundaries which fall into two categories: shock protection boundaries and arc flash protection boundaries. (Figure 5-1) Remember that only qualified persons are permitted to enter shock protection and arc flash boundaries.

Shock Protection Boundaries

The shock protection boundaries can be found by performing a shock hazard analysis. A shock hazard analysis will take into consideration the voltages to which workers will be exposed, and will detail the boundary requirements as well as the PPE necessary to perform the task safely. NFPA 70E Table 130.2(C) (Figure 5-2) can be used to calculate the shock hazard boundaries based on

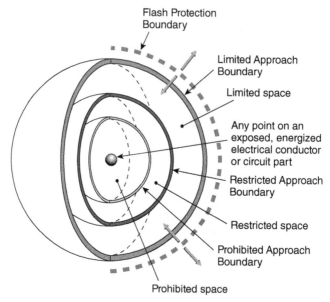

Flash Protection Boundary

Limited Approach Boundary

Limited space

Any point on an exposed, energized electrical conductor or circuit part

Restricted Approach Boundary

Restricted space

Prohibited Approach Boundary

Prohibited space

Figure 5-1 Shock and arc flash protection boundaries.
Courtesy Delmar/Cengage Learning

voltage and other variables. Some computer programs will perform this analysis, along with an arc flash analysis, and will give the results for both. Newer equipment may have the shock protection boundaries as well as the necessary PPE listed on a warning label mounted on the equipment. (Figure 5-3) It is important to keep in mind that equipment that doesn't carry a visual warning isn't necessarily free from potential hazard; older equipment will often not have this warning.

The boundaries fall into three categories: limited approach, restricted approach, and prohibited approach. The boundaries are a sphere which surrounds the energized equipment.

Limited Approach Boundary

The limited approach boundary is the minimum distance an unqualified person must stay away from exposed energized components. The "limited space" is the space between the limited approach boundary and the restricted approach boundary. (Figure 5-4) The limited

Table 130.2(C) Approach Boundaries to Energized Electrical Conductors or Circuit Parts for Shock Protection (All dimensions are distance from energized electrical conductor or circuit part to employee.)

(1)	(2)	(3)	(4)	(5)
	Limited Approach Boundary[1]		Restricted Approach Boundary[1]; Includes	
Nominal System Voltage Range, Phase to Phase[2]	Exposed Movable Conductor[3]	Exposed Fixed Circuit Part	Inadvertent Movement Adder	Prohibited Approach Boundary[1]
Less than 50	Not specified	Not specified	Not specified	Not specified
50 to 300	3.05 m (10 ft 0 in.)	1.07 m (3 ft 6 in.)	Avoid contact	Avoid contact
301 to 750	3.05 m (10 ft 0 in.)	1.07 m (3 ft 6 in.)	304.8 mm (1 ft 0 in.)	25.4 mm (0 ft 1 in.)
751 to 15 kV	3.05 m (10 ft 0 in.)	1.53 m (5 ft 0 in.)	660.4 mm (2 ft 2 in.)	177.8 mm (0 ft 7 in.)
15.1 kV to 36kV	3.05 m (10 ft 0 in.)	1.83 m (6 ft 0 in.)	787.4 mm (2 ft 7 in.)	254 mm (0 ft 10 in.)
36.1 kV to 46 kV	3.05 m (10 ft 0 in.)	2.44 m (8 ft 0 in.)	838.2 mm (2 ft 9 in.)	431.8 mm (1 ft 5 in.)
46.1 kV to 72.5 kV	3.05 m (10 ft 0 in.)	2.44 m (8 ft 0 in.)	1.0 m (3 ft 3 in.)	660 mm (2 ft 2 in.)
72.6 kV to 121 kV	3.25 m (10 ft 8 in.)	2.44 m (8 ft 0 in.)	1.29 m (3 ft 4 in.)	838 mm (2 ft 9 in.)
138 kV to 145 kV	3.36 m (11 ft 0 in.)	3.05 m (10 ft 0 in.)	1.15 m (3 ft 10 in.)	1.02 m (3 ft 4 in.)
161 kV to 169 kV	3.56 m (11 ft 8 in.)	3.56 m (11 ft 8 in.)	1.29 m (4 ft 3 in.)	1.14 m (3 ft 9 in.)
230 kV to 242 kV	3.97 m (13 ft 0 in.)	3.97 m (13 ft 0 in.)	1.71 m (5 ft 8 in.)	1.57 m (5 ft 2 in.)
345 kV to 362 kV	4.68 m (15 ft 4 in.)	4.68 m (15 ft 4 in.)	2.77 m (9 ft 2 in.)	2.79 m (8 ft 8 in.)
500 kV to 550 kV	5.8 m (19 ft 0 in.)	5.8 m (19 ft 0 in.)	3.61 m (11 ft 10 in.)	3.54 m (11 ft 4 in.)
765 kV to 800 kV	7.24 m (23 ft 9 in.)	7.24 m (23 ft 9 in.)	4.84 m (15 ft 11 in.)	4.7 m (15 ft 5 in.)

Figure 5-2 Table 130.2(C), 2009 NFPA 70E. Reprinted with permission from NFPA 70™, *National Electrical Code®*, Copyright© 2007, National Fire Protection Association, Quincy, MA 02269

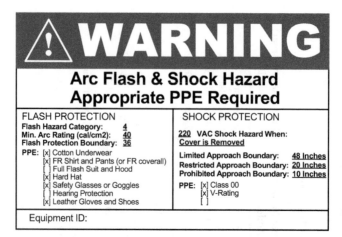

Figure 5-3 Warning label indicating arc flash boundaries and necessary PPE. *Courtesy Delmar/Cengage Learning*

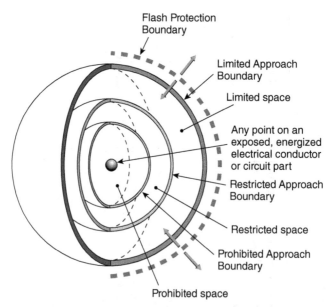

Figure 5-4 The limited approach boundary is the minimum distance an unqualified person must stay away from exposed energized parts. *Courtesy Delmar/Cengage Learning*

approach boundary is intended to keep a person who lacks skills or knowledge of the hazards far enough away that they don't inadvertently make contact with energized equipment or cause an object to fall or touch energized equipment. The unqualified person must not only keep their body out of the limited approach boundary, but also any conductive objects they may be handling. The qualified person who is in charge of the exposed energized parts has the responsibility of informing any unqualified person working near the limited approach boundary of the potential hazard, and must warn him or her not to cross the boundary. If an unqualified person must enter the limited approach boundary, a qualified person must inform the unqualified person of the potential hazards and escort them at all times.

Restricted Approach Boundary

The restricted approach boundary marks the distance that a qualified person must keep from exposed energized equipment without having both a document plan approved by the appropriate person from management and the necessary PPE. (Figure 5-5) Any part of the qualified person's body that will cross, or may cross, must be wearing PPE, which will insulate them from the voltage and energy level that is present. An unqualified person should never cross the restricted approach boundary.

Prohibited Approach Boundary

The prohibited approach boundary marks the area around exposed energized equipment which is considered to be as hazardous as actually being in contact with the energized components. (Figure 5-6) A qualified person crossing the prohibited approach boundary must meet the following requirements:

- Have the training necessary to work on energized parts with the voltage and energy level he or she will be exposed to;
- Have a documented plan, signed by the appropriate person from management, justifying the need to work on the equipment while it is in an energized state;

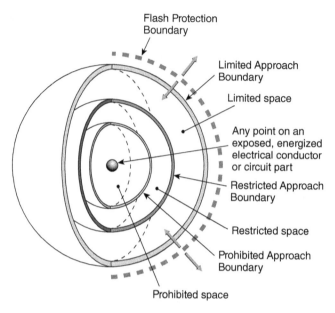

Flash Protection
Boundary

Limited Approach
Boundary

Limited space

Any point on an
exposed, energized
electrical conductor
or circuit part

Restricted Approach
Boundary

Restricted space

Prohibited Approach
Boundary

Prohibited space

Figure 5-5 The restricted approach boundary is the
distance away from exposed energized equipment that
a qualified person must keep away without having a
document plan approved by the appropriate person from
management and wearing the necessary PPE. *Courtesy
Delmar/Cengage Learning*

- Have performed a risk analysis which is signed by the appropriate
  person from management;
- Use the appropriate PPE and insulated tools rated for the voltage
  and energy level to which he or she will be exposed.

Arc Flash Boundary

The arc flash boundary marks the distance from energized electrical
equipment within which a person could receive a second-degree burn
in the case of an arcing incident. (Figure 5-7) This boundary is found
by performing an arc flash hazard analysis. This analysis is a complicated procedure which is typically performed by electrical engineers
and is beyond the scope of this text. The arc flash hazard analysis will

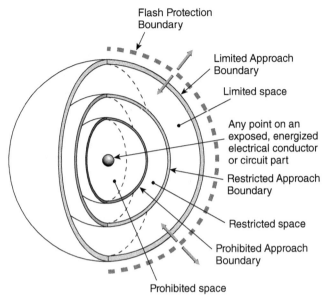

Figure 5-6 The prohibited approach boundary is the distance away from exposed energized equipment that is considered to be the same as coming into contact with the energized parts. *Courtesy Delmar/ Cengage Learning*

determine the arc flash boundary as well as the necessary PPE. When an arc flash hazard analysis has been performed and a potential hazard has been found, the arc flash boundary distance as well as the necessary PPE should be printed on a label mounted on the equipment. It is important to remember that the absence of an arc flash label doesn't mean the equipment is free from arc flash hazard.

The NFPA 70E states that if an arc flash hazard analysis hasn't been performed then the arc flash boundary shall be four feet, on systems between 50 and 600 volts, with an available bolted fault current of 50,000 amps or less, and a clearing time of 2 cycles or less. (More details on this scenario can be found in Section 130.3(A)(1) of the NFPA 70E). Whenever you are working around energized equipment that hasn't had an arc flash hazard analysis performed, remember that the minimum arc flash boundary is four feet.

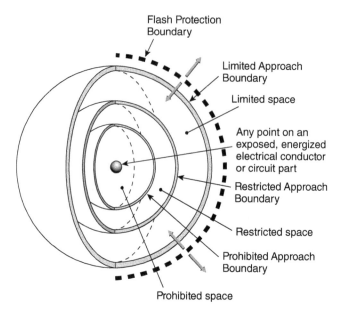

Flash Protection Boundary

Limited Approach Boundary

Limited space

Any point on an exposed, energized electrical conductor or circuit part

Restricted Approach Boundary

Restricted space

Prohibited Approach Boundary

Prohibited space

Figure 5-7 The arc flash boundary is the distance from energized electrical equipment where a person could receive a second-degree burn in the case of an arcing incident. *Courtesy Delmar/Cengage Learning*

NFPA 70E Tables

NFPA 70E permits applying the requirements of 130.7(C)(9), 130.7(C)(10), 130.7(C)(11), and the associated tables in lieu of a detailed incident energy analysis for determining the appropriate PPE. If using this method, it is important to understand that the tables are based on typical scenarios, and some situations will not fall under the parameters of the table.

- Table 130.7(C)(9) lists tasks that may be performed and the specific types of energized equipment a task may be performed on. It lists the hazard/risk category associated with each task and whether insulated gloves and/or insulated hand tools will be required. (Figure 5-8)
- Table 130.7(C)(10) is titled Protective Clothing and Personal Protective Equipment. This table lists the hazard/risk categories and details the clothing and PPE that are required for each category. (Figure 5-9)
- Table 130.7(C)(11) is titled Protective Clothing Characteristics. This table gives additional information about protective clothing as it pertains to each hazard/risk category. (Figure 5-10)

Table 130.7 (C)(9) Hazard/Risk Category Classifications and Use of Rubber Insulating Gloves and Insulated and Insulating Hand Tools

Tasks Performed on Energized Equipment	Hazard/Risk Category	Rubber Insulating Gloves	Insulated and Insulating Hand Tools
Panelboards or Other Equipment Rated 240 V and Below — Note 1			
Perform infrared thermography and other non-contact inspections outside the restricted approach boundary	0	N	N
Circuit breaker (CB) or fused switch operation with covers on	0	N	N
CB or fused switch operation with covers off	0	N	N
Work on energized electrical conductors and circuit parts, including voltage testing	1	Y	Y
Remove/install CBs or fused switches	1	Y	Y
Removal of bolted covers (to expose bare, energized electrical conductors and circuit parts)	1	N	N
Opening hinged covers (to expose bare, energized electrical conductors and circuit parts)	0	N	N
Work on energized electrical conductors and circuit parts of utilization equipment fed directly by a branch circuit of the panelboard	1	Y	Y
Panelboards or Switchboards Rated >240 V and up to 600 V (with molded case or insulated case circuit breakers) — Note 1			
Perform infrared thermography and other non-contact inspections outside the restricted approach boundary	1	N	N
CB or fused switch operation with covers on	0	N	N

Figure 5-8 Table 130.7(C)(9), 2009 NFPA 70E. Reprinted with permission from NFPA 70™, *National Electrical Code®*, Copyright© 2007, National Fire Protection Association, Quincy, MA 02269

Table 130.7(C)(10) Protective Clothing and PPE

Hazard/Risk Category	Protective Clothing and PPE

Hazard/Risk Category 0
Protective Clothing, Nonmelting
   (according to ASTM F 1506-00)
   or Untreated Natural Fiber
	Shirt (long sleeve)
	Pants (long)

FR Protective Equipment	Safety glasses or safety goggles (SR) Hearing protection (ear canal inserts) Leather gloves (AN) (Note 2)

Hazard/Risk Category 1
FR Clothing. Minimum Arc Rating
   of 4 (Note 1)
	Arc-rated long-sleeve shirt (Note 3)
	Arc-rated pants (Note 3)
	Arc-rated coverall (Note 4)
	Arc-rated face shield or arc flash suit hood
	   (Note 7)
	Arc-rated jacket, parka, or rainwear (AN)

FR Protective Equipment	Hard hat Safety glasses or safety goggles (SR) Hearing protection (ear canal inserts) Leather gloves (Note 2) Leather work shoes (AN)

Hazard/Risk Category 2
FR Clothing, Minimum Arc Rating
   of 8 (Note 1)
	Arc-rated long-sleeve shirt (Note 5)
	Arc-rated pants (Note 5)
	Arc-rated coverall (Note 6)
	Arc-rated face shield or arc flash suit hood
	   (Note 7)
	Arc rated jacket, parka, or rainwear (AN)

FR Protective Equipment	Hard hat Safety glasses or safety goggles (SR) Hearing protection (ear canal inserts) Leather gloves (Note 2) Leather work shoes
Hazard/Risk Category 2* FR Clothing, Minimum Arc Rating of 8 (Note 1)	Arc-rated long-sleeve shirt (Note 5) Arc-rated pants (Note 5) Arc-rated coverall (Note 6) Arc-rated arc flash suit hood (Note 10) Arc-rated jacket, parka, or rainwear (AN)
FR Protective Equipment	Hard hat Safety glasses or safety goggles (SR) Hearing protection (ear canal inserts) Leather gloves (Note 2) Leather work shoes

Figure 5-9 Table 130.7(C)(10), 2009 NFPA 70E. Reprinted with permission from NFPA 70™, *National Electrical Code*®, Copyright© 2007, National Fire Protection Association, Quincy, MA 02269

Hazard/Risk Category	Protective Clothing and PPE
Hazard/Risk Category 3 **FR** Clothing, Minimum Arc Rating of 25 (Note 1)	Arc-rated long-sleeve shirt (AR) (Note 8) Arc-rated pants (AR) (Note 8) Arc-rated coverall (AR) (Note 8) Arc-rated arc flash suit jacket (AR) (Note 8) Arc-rated arc flash suit pants (AR) (Note 8) Arc-rated arc flash suit hood (Note 8) Arc-rated jacket, parka, or rainwear (AN)
FR Protective Equipment	Hard hat FR hard hat liner (AR) Safety glasses or safety goggles (SR) Hearing protection (ear canal inserts) Arc-rated gloves (Note 2) Leather work shoes
Hazard/Risk Category 4 FR Clothing, Minimum Arc Rating of 40 (Note 1)	Arc-rated long-sleeve shirt (AR) (Note 9) Arc-rated pants (AR) (Note 9) Arc-rated coverall (AR) (Note 9) Arc-rated arc flash suit jacket (AR) (Note 9) Arc-rated arc flash suit pants (AR) (Note 9) Arc-rated arc flash suit hood (Note 9) Arc-rated jacket, parka, or rainwear (AN)
FR Protective Equipment	Hard hat FR hard hat liner (AR) Safety glasses or safety goggles (SR) Hearing protection (ear canal inserts) Arc-rated gloves (Note 2) Leather work shoes

AN = As needed (optional)
AR = As required
SR = Selection required
Notes:

1. See Table 130.7(C)(11). Arc rating for a garment or system of garments is expressed in cal/cm^2.
2. If rubber insulating gloves with leather protectors are required by Table 130.7(C)(9), additional leather or arc-rated gloves are not required. The combination of rubber insulating gloves with leather protectors satisfies the arc flash protection requirement.
3. The FR shirt and pants used for Hazard/ Risk Category 1 shall have a minimum arc rating of 4.
4. Alternate is to use FR coveralls (minimum arc rating of 4) instead of FR shirt and FR pants.
5. FR shirt and FR pants used for Hazard/ Risk Category 2 shall have a minimum arc rating of 8.
6. Alternate is to use FR coveralls (minimum arc rating of 8) instead of FR shirt and FR pants.
7. A face shield with a minimum arc rating of 4 for Hazard/Risk Category 1 or a minimum arc rating of 8 for Hazard/Risk Category 2, with wrap-around guarding to protect not only the face, but also the forehead. ears, and neck (or, alternatively, an arc-rated arc flash suit hood), is required.
8. An alternate is to use a total FR clothing system and hood, which shall have a minimum arc rating of 25 for Hazard/Risk Category 3.
9. The total clothing system consisting of FR shirt and pants and/or FR coveralls and/or arc flash coat and pants and hood shall have a minimum arc rating of 40 for Hazard/Risk Category 4.
10. Alternate is to use a face shield with a minimum arc rating of 8 and a balaclava (sock hood) with a minimum arc rating of 8 and which covers the face, head and neck except for the eye and nose areas.

Figure 5-9 *Continued.*

Table 130.7(C)(11) Protective Clothing Characteristics

Hazard/ Risk Category	Clothing Description	Required Minimum Arc Rating of PPE [J/cm²(cal/cm²)]
0	Nonmelting, flammable materials (i.e.. un-treated cotton, wool, rayon, or silk, or blends of these materials) with a fabric weight at least 4.5 oz/yd²	N/A
1	Arc-rated FR shirt and FR pants or FR coverall	16.74(4)
2	Arc-rated FR shirt and FR pants or FR coverall	33.47 (8)
3	Arc-rated FR shirt and pants or FR coverall, and arc flash suit selected so that the system arc rating meets the required minimum	104.6 (25)
4	Arc-rated FR shirt and pants or FR coverall, and arc flash suit selected so that the system arc rating meets the required minimum	167.36 (40)

Figure 5-10 Table 130.7(C)(11), 2009 NFPA 70E. Reprinted with permission from NFPA 70™, *National Electrical Code®*, Copyright© 2007, National Fire Protection Association, Quincy, MA 02269

Working on Energized Equipment

When it is truly justified to work on energized equipment, there are several dangers that must be understood, steps to be followed, and appropriate PPE to be worn.

Energized Work Permit

Most work performed on an energized circuit will require a written energized work permit. The permit must be filled out before the work is to commence. The one exception to this is diagnosing and testing, provided the testing is being performed by a qualified person wearing the appropriate PPE.

According to the NFPA 70E, an energized work permit shall contain the following:

- Description of the circuit and equipment to be worked on;
- Description of the work to be completed;
- Justification for the work to be performed while the equipment is energized;

- Description of the safe work practices that will be employed;
- Results of the shock hazard analysis;
- Shock protection boundaries;
- Results of an arc flash hazard analysis;
- Arc flash boundary;
- Necessary PPE;
- Means of restricting access to unqualified people while the work is being performed;
- Evidence of a job briefing that included a discussion of the hazards;
- Energized work approval (signature from authorizing individual).

Establishing a Procedure

It is important to have a procedure in place for working on energized equipment. This will help to ensure that all steps are followed and all precautions taken. The procedure will vary based on the type of task being performed.

The following are examples of a few items which could be part of a procedure to troubleshoot equipment:

- Based on the equipment and situation, identify the expected voltages;
- Don the appropriate PPE based on the hazard present (safety glasses, hard hat, voltage-rated gloves, ear protection, arc flash suit, voltage-rated mat, etc.);
- Open the equipment to expose the components being tested;
- If needed, insulate any energized equipment that may create a potential shock hazard, using a voltage-rated insulating blanket;
- If needed, block any components that may create an arc flash hazard using an arc suppression blanket;
- Verify that the meter is working properly on a known voltage source;
- Take the measurements to troubleshoot the circuit;
- Verify that the meter is still working on a known voltage source;
- Remove all safety components installed (voltage-rated blanket, arc suppression blanket, etc.);
- Replace doors or covers.

Awareness

Awareness includes self-awareness as well as cognizance of what is going on around you. Many accidents have been caused by a person who is tired, under the influence of a legal or illegal drug, or simply not paying attention.

Alertness

Be sure that you are alert and have a complete understanding of the task at hand, as well as what is going on around you. This includes identifying all the energized components that could create a hazard; being aware of any materials that could move or fall onto you or into energized equipment; and any other work going on around you that may have an impact on what you are doing or create a hazard. Pay attention to where all parts of your body are, where they are moving to, and what they could come into contact with. It is easy to be so focused on what you are doing with one hand that the other will inadvertently touch an energized piece of equipment. Always remember how dangerous working with energized equipment can be. When workers become complacent, they begin to cut corners, make mistakes, and get careless—a recipe for disaster.

Anticipating failure or problems is another important part of being alert. This is a key component of the electrical permit and shock hazard analysis. It is important to think about situations that may arise and problems that could be encountered before performing the work. That way it won't be a surprise if a problem occurs, and the worker will have already planned on how to deal with it.

Impairedness

Being impaired is often thought of as being under the influence of alcohol or an illegal drug. It should go without saying that a person should never work on or near energized equipment—or even be on a construction site—under the influence of alcohol or illegal drugs. What people tend to forget is that legal drugs can also have many negative effects that will create a hazard on the job. They might make a person dizzy, inattentive, or tired, or slow one's reflexes. If you are taking any drugs—either prescription or non-prescription—ask your doctor about any negative effects that may create a hazard before you begin working on a construction site, and especially around energized electrical equipment.

Blind Reaching

Blind reaching is reaching into an area that a person can't see, and is prohibited by the NFPA 70E, as you may come into contact with energized components. If you can't see where your hands are going, you have no business putting them there.

Working Conditions

Cleanliness is an important part of every job, even more so when one is working around energized electrical equipment. Parts and tools lying around can cause a person to trip or stumble and touch an energized part or cause an arcing incident. Clutter may also get in the way of a quick retreat in the event of an arcing incident. The list of dangers associated with a messy or cluttered workplace could go on and on; by keeping your work area clean, your job will be easier and safer.

Illumination is a vital part of working on energized equipment. Dim or dark working conditions will create a hazard, as you won't be able to see what you are doing or the dangers that may be present. The NFPA 70E addresses this hazard and prohibits working in the limited approach boundary without proper illumination. This may require temporary lights to be brought in.

Personal Protective Equipment

When working around energized electrical equipment a person must wear the appropriate PPE necessary to protect against electrical shock and arcing incidents.

Electrical Shock

To protect against electrical shock, there must be a barrier between the worker and energized electrical parts. Having the proper training and using the appropriate PPE can allow the qualified person to work safely on energized equipment.

Rubber Insulated Personal Protective Equipment

Rubber is found in most of the PPE designed to protect against electrical shock. Rubber PPE is designated by type (TYPE I and TYPE II) and class (00, 0, 1, 2, 3, 4) Type I rubber PPE is made of a rubber which is susceptible to ozone checking (cracking). Exposure to ozone creates the cracks found on rubber products, such as are often found on rubber bands, car tires, gaskets, etc. Type II rubber PPE is made of synthetic rubber and is more resistant to ozone and less likely to crack. The classes that rubber PPE is divided into will indicate the voltage that particular PPE has been designed to handle. (Figure 5-11)

Rubber PPE is often available in different colors to meet the various needs of employers. Some rubber PPE will be manufactured in multiple colors. Gloves, for example, are often available with one color on the inside and another on the outside. The advantage of this is that it makes it easier to find cuts, holes, and scratches while the glove is

Insulating Rubber Classification	
Type I	Non-Ozone Resistant
Type II	Ozone Resistant
Class	Maximum AC Voltage (RMS)
00	500V
0	1,000V
1	7,500V
2	17,000V
3	26,500V
4	36,000V

Figure 5-11 Insulating rubber classification chart.
Courtesy Delmar/ Cengage Learning

inflated. The second color will often show through any damaged spots or imperfections in the rubber.

Visual inspection and electrical testing is an important part of the proper care of rubber PPE and is required by OSHA. Storage and cleaning of rubber PPE is also extremely important to ensure that it will provide the intended protection. Always follow the manufacturer's instructions for proper care and storage of PPE.

Voltage-Rated Gloves

Insulated gloves are the first line of defense against electrical shock. There are two parts to the typical set of voltage-rated gloves: the rubber insulating glove and the glove protector. (Figure 5-12) The insulating glove is made of a rubber material which offers protection against the rated voltage, and the leather glove protector protects the gloves from punctures and cuts during use. Voltage-rated gloves come in different voltage ratings and sizes. Although it may seem logical to use the gloves with the highest voltage rating for all tasks, these may be unnecessarily cumbersome. The higher the voltage rating, the thicker the glove and the less dexterity a person will have. Proper fit is also a key part of having a voltage-rated glove that is easy to manipulate.

Figure 5-12 Voltage-rated gloves, leather protector, and case. *Courtesy of Salisbury by Honeywell*

Gloves that are too small will restrict movement, while gloves that are too big will be sloppy and cumbersome. Proper fit is found by measuring around the palm of the dominant hand at its largest circumference. (Figure 5-13) Gloves that are cumbersome or restrict movement are dangerous and can lead to accidents.

Rubber insulating gloves must be properly cared for and tested to ensure they maintain the level of protection intended. Before each use and after any incident that is suspected of causing damage, a glove must be visually inspected inside and out for nicks, cuts, punctures, ozone checking, embedded objects, and chemical deterioration (softness, hardness, swelling, or stickiness). A daily inflation test must also be performed. (Figure 5-14) During an inflation test, the glove is filled with air and checked for leaks by listening for escaping air. While the glove is inflated, it is very easy to see any physical damage that may have occurred to the glove, such as nicks, scratches, etc. The inflation test should be performed with the glove right-side-out as well as inside-out. A glove inflator can be used to inflate the glove, or the cuff can be rolled, trapping air in the glove. When inflating gloves, it is

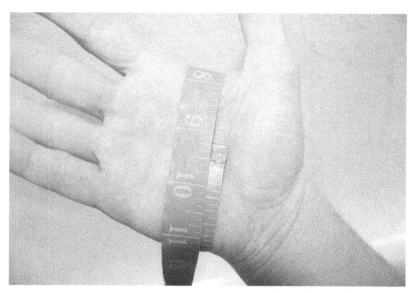

Figure 5-13 Proper fit for voltage-rated gloves is found by measuring the largest part of the palm on a person's dominant hand. *Courtesy Delmar/ Cengage Learning*

Figure 5-14 A glove inflator will inflate a voltage-rated glove so it can be inspected for damage. *Courtesy of Salisbury by Honeywell*

important to not overinflate. Be sure to read the manufacturer's instructions on inflation and inspection for the particular gloves you are using. If damage is found or suspected, the gloves must be taken out of service immediately and either destroyed or sent off for electrical testing.

Many chemicals, oils, and lotions will cause deterioration of rubber gloves and should be avoided. Petroleum-based products such as gasoline, oil, hydraulic fluid, and solvents are particularly harmful to rubber products. If there is contact with a substance which could cause deterioration it should be washed off immediately. Gloves should be cleaned with water and a mild soap, rinsed thoroughly, and allowed to air-dry.

A worker shouldn't be wearing rings or jewelry at work, but it is particularly important to not wear any watches or jewelry under rubber insulating gloves, as they can cause damage.

Gloves should be stored in a protective bag when not in use: only store one pair of gloves per bag. Insert the gloves into the bag with the cuff side down—fingers up—and hang the bag. NEVER fold the gloves or store items on top of them, as damage will result. Store the gloves in a cool, dry location away from sunlight and sources of ozone.

Gloves must be electrically tested at least once every six months. Gloves that have been in storage for over 12 months must also be tested prior to use.

Leather protectors must also be inspected for damage. Pieces of wire, metal shavings, etc. that are imbedded in the leather protector can cause damage to the rubber glove and must be removed. Also look for tears, cuts, open seams, and oil spots.

Insulating Sleeves

Insulating sleeves, sometimes called voltage-rated sleeves, are used to protect the arms from accidental contact while reaching into equipment with energized components. (Figure 5-15) Before each use, sleeves should be visually inspected inside and out for any possible damage, just as voltage-rated gloves are. Voltage-rated sleeves must be electrically tested at least every 12 months or after any incident where damage may have occurred. If damage is found or suspected, the sleeves must be taken out of service immediately and sent off for electrical testing.

Figure 5-15 A worker wearing voltage-rated sleeves and gloves. *Courtesy of Salisbury by Honeywell*

Insulated sleeves must be stored in a cool, dry place away from sunlight, like voltage-rated gloves. It is also important that the sleeves are not folded over or stored underneath anything. They should be placed in their bag and hung to prevent any damage during storage.

Voltage-Rated Blankets

Voltage-rated blankets are used to provide a barrier between energized equipment and the worker. (Figure 5-16) For instance, when working in a piece of switchgear with energized parts, the insulated blanket can be hung or wrapped around energized components to help prevent unintentional contact. Other PPE, such as voltage-rated gloves and sleeves, will still be required to provide additional protection.

Before each use, a voltage-rated blanket must be inspected for embedded debris, punctures, or any other damage. Voltage-rated blankets must be tested at least every 12 months.

Voltage-rated blankets must be rolled up and stored in their canisters in a cool, dry place. Do not fold the blanket up or store items on it, as resulting damage can jeopardize the insulation.

Voltage-Rated Mats

Voltage-rated mats provide a surface for the worker to stand on when working on energized electrical components. (Figure 5-17) It is intended to provide an additional barrier between the worker and a potential electrical shock. Caution must be used, as any crack or piece of debris will compromise the integrity of the mat, which is typically used as supplemental protection in addition to other insulating PPE.

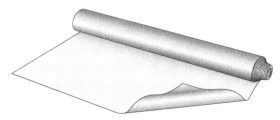

Figure 5-16 Voltage-rated blanket. *Courtesy of Salisbury by Honeywell*

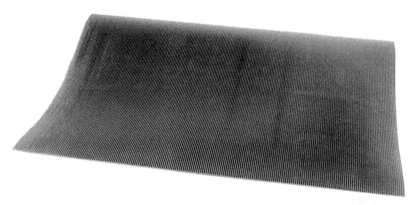

Figure 5-17 Voltage-rated mat. *Courtesy of Salisbury by Honeywell*

It is important to follow the manufacturer's recommendations on use, inspection, and testing of these mats.

Insulated Footwear

Work boots are available with an electrical hazard (EH) rating. (Figure 5-18) They are typically 600 volt-rated and will offer protection against electric current passing through the sole of the foot. These boots are a great idea for the electrical worker, as they add one more level of protection against electrical shock. The problem is that they can only guarantee this protection when taken out of the box the first time. After wearing them around, especially on a construction site, nails, glass, rocks, and everyday wear have the potential to damage the insulation. Add the moisture from a perspiring foot or from stepping in water, and it is easy to see how there could be a potential current path. Because of this, the NFPA 70E does not recognize the insulated sole on work boots as providing primary step and touch protection. If this type of protection is necessary, then insulated

EH

Figure 5-18 Boots are available with an EH rating. *Courtesy Delmar/Cengage Learning*

Figure 5-19 Voltage-rated overshoes. *Courtesy of Salisbury by Honeywell*

overshoes must be used. Insulated overshoes must be properly taken care of and inspected prior to each use to ensure the rated protection. (Figure 5-19)

Even though the insulation rating on EH boots shouldn't be relied upon as the only level of protection against electrical shock, wearing them is a good idea; they just may provide that extra level of insulation necessary to save you from a lethal shock.

Insulated Tools

Any tool that is going to be used within the limited approach boundary that is either intended to, or may accidentally, come into contact with energized parts must have voltage-rated insulation on the tool. Many people think that the plastic coating on pliers, a screwdriver, etc. is an insulator that will protect them from an electrical shock; this is simply not true. The plastic handle found on most tools is intended to make the tool more comfortable and easier to use, and does not offer protection against electrical shock. (Figure 5-20) There are tools on the market that do have voltage-rated insulation. (Figure 5-21) Voltage-rated tools will indicate the maximum voltage they protect against. Care must be taken to never exceed the maximum voltage for which a tool is rated. Care must also be taken to not damage the insulation in any way, or the tool will not offer protection from the intended voltage. Insulated tools should be inspected for damage prior to each use; if any damage is found, the tool must be replaced.

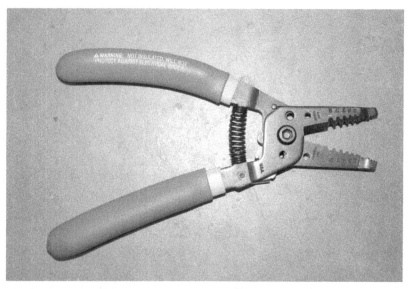

Figure 5-20 Most hand tools have handles designed for comfort and do not offer any voltage protection. *Courtesy Delmar/Cengage Learning*

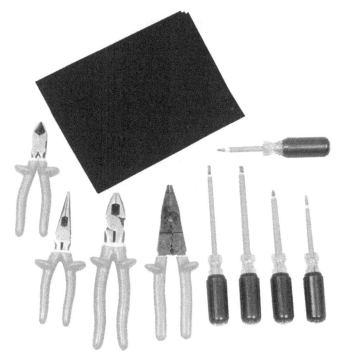

Figure 5-21 Voltage-rated hand tools. *Courtesy of Salisbury by Honeywell*

Arcing Flash/Blast Personal Protective Equipment

When working on energized equipment, the worker must not only protect himself against electrical shock, but also against an arcing incident. To protect a worker from arcing incidents, the appropriate PPE must be worn to provide a barrier between the arc flash/blast and the worker.

A label on equipment that has undergone an arc flash hazard analysis may indicate the arc flash hazard as well as the necessary PPE. The NFPA 70E has tables which indicate the appropriate level of PPE based on the job being performed and the arc flash hazard. Be sure to read the label or check the NFPA tables to ensure that the correct PPE is being worn.

Safety Glasses

When working on energized equipment, OSHA-approved safety glasses with side shields must be worn. In the case of an arcing incident, whether it is major or minor, safety glasses may save your eyesight. Remember that safety glasses must be worn underneath face shields and arc flash hoods.

Flame-Resistant Clothing

When working around equipment that has the potential for an arcing incident, it is imperative to wear the proper clothing. (Figure 5-22) Exposure to the high temperatures present with an arc flash will cause synthetic fibers such as nylon and polyester to melt, while others will ignite and continue to burn after the arcing incident has ended. Many people have been seriously burned by synthetic fibers that melted into their skin during an arcing incident. Cotton and some other natural fibers won't melt when exposed to high temperatures, but they may catch on fire, which will also cause serious burns. Flame-resistant clothing either won't burn or will self-extinguish after the flame has been removed. Note: tight-fitting clothing should be avoided, as it carries a greater risk of being ripped or blown off in an arcing incident.

Figure 5-22 Flame-resistant clothing will have an arc rating which indicates the amount of protection offered. *Courtesy Delmar/Cengage Learning*

As the arc flash hazard increases, so does the requirement for flame-resistant clothing. Flame-resistant clothing is rated by its arc thermal performance value (ATPV), which will be identified on the garment. The clothing may also list the maximum hazard/risk category the clothing is safe for. It is important to remember that any undergarments worn must also be non-melting. Socks, underwear, and bras must be 100 percent cotton or made of a flame-resistant material.

Clothing that is made of flame-resistant materials has very specific washing and drying requirements. Chemicals and high temperatures can cause damage which will jeopardize a garment's ability to offer protection. The number of times a garment is washed is another factor: most flame-resistant clothing will have a maximum number of times that it can be washed before it must be replaced. Be sure to read the manufacturers' recommendations for the cleaning of flame-resistant garments.

Arc-Rated Face Shield

A face shield will offer protection from heat and debris which could damage the face and ears during an arcing incident. (Figure 5-23) The shield must be designed to protect the face, forehead, ears,

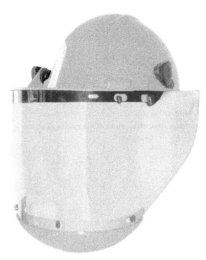

Figure 5-23 Arc-rated face shield. *Courtesy of Salisbury by Honeywell*

and neck, in addition to being arc flash-rated. A regular face shield designed only to protect against flying objects will melt and catch fire in the case of an arcing incident.

Balaclava

A balaclava, which looks like a face mask, is worn to protect a person's head, neck, ears, and face. It must have the correct flame resistance ratings, and be designed for use where there is an arc flash hazard.

Hard Hat

A Class E hard hat is required when working on equipment with hazard/risk categories of 1, 2, 3, or 4. In the event of an arc blast, debris will be projected at extremely high velocities. A hard hat will offer protection against flying or falling objects which could cause head injuries. Higher-risk categories require a flame-resistant hard hat liner.

Ear Protection

Arcing incidents produce extremely loud noise which is damaging to the ears. Ear protection is required for all hazard/risk categories. Hearing protection inserted into the ear canal works best with the other

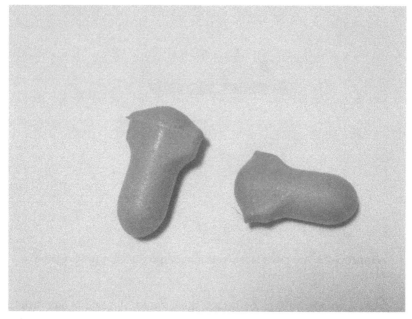

Figure 5-24 Foam ear plugs can be inserted into the ear canal and worn under a flash hood. *Courtesy Delmar/Cengage Learning*

PPE necessary when there is an arc flash hazard. (Figure 5-24) Ear plugs are now available that will allow low level sounds such as conversations, but will block out loud sudden noises such as an arc blast.

Work Boots

Good work boots are a necessity on the job site. They provide protection for the foot and can relieve stress on the back. When working around an arc flash hazard, the NFPA 70E requires the footwear to be leather. A person wearing boots or shoes that are made of non-leather material which may burn or melt can suffer serious injuries.

Gloves

Hands are extremely vulnerable and are often closest to the action. When working near equipment that is energized, gloves are a must. If there is a shock hazard, voltage-rated gloves must be used—these will also provide protection against an arc flash. Where there is an arc flash hazard but no electrical shock hazard, NFPA 70E requires leather or flame-resistant gloves.

Arc Flash Suit

An arc flash suit consists of arc-rated pants or bibs, jacket, and hood. (Figure 5-25) The hood often has a hard hat built in. Arc flash suits are available in several levels of protection to meet the different levels of arc flash hazard. Remember that an arc flash suit alone doesn't offer all the protection you need. Flame-resistant clothing, safety glasses, leather work boots, leather gloves, hard hat (if not built into the arc flash suit), and ear protection are still to be worn in addition to the arc flash suit.

Figure 5-25 Worker wearing arc flash suit. *Courtesy of Salisbury by Honeywell*

Arc Suppression Blanket

Arc suppression blankets are intended to redirect the explosive energy of an arc flash or blast. (Figure 5-26) It is important to keep in mind that they do not provide arc flash/blast protection; that is provided by the appropriate clothing, flash suit, etc. The effects of an arcing incident are very unpredictable and an arc suppression blanket is intended to redirect some of the effects of the arc flash. It is also important to note that arc suppression blankets do not provide the shock protection of an insulating blanket. Be sure to follow the manufacturer's instructions on installation, inspection, and storage of the blanket.

Although there is an abundance of PPE on the market to protect the worker in various scenarios where an arc flash hazard exists, the best way to stay safe is to disconnect the power and perform the work when the equipment is de-energized. Remember that OSHA

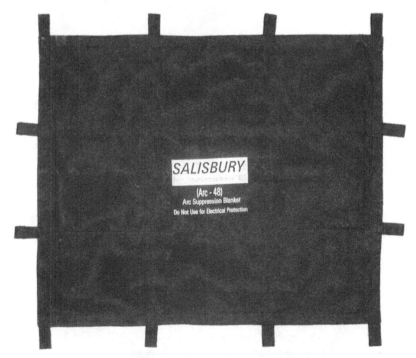

Figure 5-26 Arc suppression blanket. *Courtesy of Salisbury by Honeywell*

and the NFPA 70E only permit working on energized equipment in cases where disconnecting the power will create a greater hazard or is infeasible due to equipment design or operational limitations.

Safety Signs, Barricades, and Attendants

It is the responsibility of the qualified worker to ensure that unqualified persons don't enter work areas where shock or arc flash hazards exist due to exposed energized equipment. While performing work on energized electrical equipment, it may be necessary to install safety signs, barricades, or even have attendants on hand to keep unqualified workers out of the area.

Summary

- Work should never be performed on energized electrical equipment unless disconnecting the power will create a greater hazard, or if it is infeasible due to equipment design or operational limitations.
- The limited approach boundary is the minimum distance an unqualified person must stay away from exposed energized parts.
- The restricted approach boundary is the minimum distance a qualified person must stay away from exposed energized parts without wearing the appropriate PPE.
- The prohibited approach boundary marks the area around exposed energized equipment which is considered as dangerous as coming into contact with the energized equipment itself.
- The arc flash boundary is the distance from energized electrical equipment where a person will receive a second-degree burn in the case of an arcing incident.
- The NFPA 70E has several tables which can help determine the hazard/risk category of a task, as well as the appropriate PPE required.

- When working on energized electrical equipment, the appropriate PPE must be worn/used to protect against an electrical shock.
  - Voltage-rated gloves
  - Insulating sleeves
  - Voltage-rated blankets
  - Voltage-rated mats
  - Insulated footwear
  - Insulated tools
- Tools used on energized electrical equipment must have insulation that is voltage-rated and must be inspected prior to each use.
- When working on or near exposed energized electrical equipment that poses an arc flash hazard, the appropriate PPE must be worn.
  - Safety glasses
  - Flame-resistant clothing
  - Arc-rated face shield
  - Balaclava
  - Hard hat
  - Ear protection
  - Work boots
  - Gloves
  - Arc flash suit
  - Arc suppression blanket
- All PPE must be inspected for damage prior to each use to ensure it will provide the intended protection.

Review Questions

1. List three examples where disconnecting the power to work on equipment may create an additional or increased hazard.
2. Which approach boundary is considered the same as coming into contact with energized components?
3. Which NFPA 70E table lists the hazard/risk category associated with a specific task?
4. List five things that must be included in an energized work permit.

5. Why do rubber gloves often have one color on the inside and another on the outside?

6. How often do voltage-rated gloves need to be sent for testing?

7. Footwear that is insulated to prevent electrical shock is marked
_____.

8. List five items of PPE that are used to protect against electrical shock.

9. List five items of PPE that are used to protect against an arcing incident.

6

Lockout-Tagout

Objectives

- Describe the need for lockout-tagout (LOTO) procedures

- Describe how an employer's safety plan impacts lockout-tagout procedures

- List items that should be included in a lockout-tagout procedure

- Identify various lockout-tagout devices

- Describe various lockout-tagout devices

- Describe the steps in a lockout-tagout procedure

Introduction

The general rule laid down by OSHA and the NFPA 70E requires that all potential energy be removed from equipment before it is worked on. In addition to removing all energy, equipment must be prevented from being energized while the work is being performed. (Figure 6-1)

Lockout-tagout (LOTO) procedures are intended to ensure that all potential energy sources have been removed from the equipment that is to be worked on, and to prevent the equipment from being un-intentionally started up or having stored energy released.

Lockout-tagout procedures must be used whenever there is potential danger from electricity or any other possible energy source. (Figure 6-2) The following are a few examples of scenarios where LOTO is required:

- Constructing a circuit which could become energized;
- Installing devices, equipment, or machinery ;
- Setting up or adjusting equipment or machinery;

Figure 6-1 Locking out and tagging out equipment will notify people that the equipment has been intentionally shut off and will prevent the equipment from being turned back on. *Courtesy Delmar/Cengage Learning*

Figure 6-2 Always verify the absence of voltage before terminating devices. *Courtesy Delmar/Cengage Learning*

- Inspecting equipment or machinery;
- Repairing equipment or machinery;
- Modifying equipment or machinery;
- Replacing devices or equipment.

LOTO and the Safety Plan

An employer's safety plan is required to include a procedure for LOTO. This procedure must be in writing and clearly describe the steps and actions required. Lockout-tagout procedures may vary from one employer to another and from one piece of equipment to another. For example, the LOTO procedures for disconnecting a large piece of machinery, with several sources of energy, in an industrial facility will differ from the LOTO plan for disconnecting a circuit that feeds a light fixture. Plans will also vary from employer to employer.

OSHA and the NFPA 70E have general LOTO requirements which must be a part of an employer's LOTO procedures. The procedure itself is left up to the employer, so they can customize it to best fit their needs. It is also important to remember that you may need to work together with other trades or companies whose lockout-tagout procedures may be different than yours. For example, if you are going to service or repair a piece of equipment that has failed or been damaged, it is very likely that there will be other trades working on the equipment as well. The other trades and the facility where the equipment is located may have their own LOTO requirements. Before work is to commence, this should be discussed so that everyone is using the same plan.

Lockout-tagout training must be a part of the employer's safety program. The employees must be trained on the LOTO procedures so they don't put themselves or others at risk. Training may be required on several procedures, as the LOTO requirements may vary by the type of equipment the work is to be performed on and the task at hand. It is also important to remember that retraining may be necessary if an employee hasn't performed a given task in a while. As with all safety requirements, the employer has the responsibility to provide the training, but the final responsibility falls on the employee to follow the procedures.

Lockout-Tagout

There are several aspects of LOTO. The following is a list of items which should be included in a LOTO procedure:

- Planning;
- Notification of all persons associated with the machine or equipment that is to be shut down;
- Shutting down of equipment;
- Removal of all energy sources;
- Application of locks and tags;
- Verification of equipment isolation and the absence of energy sources;
- If applicable, the grounding of all conductors which could become energized.

Planning

Planning a shutdown includes investigating the energy sources that need to be disconnected, discharged, blocked, etc., as well as looking at the effects of the shutdown, and may also include a written plan.

Investigation

Some shutdowns are minor and don't require much investigation. An example could be disconnecting the power to a light. Removing the energy source from the light could be as simple as turning off the switch or circuit breaker. Other shutdowns may be very complex and will necessitate looking at blueprints, schematic diagrams, wiring diagrams, or other drawings to locate all the energy supplies and figure out how to safely isolate the equipment. Machinery in large industrial facilities and assembly lines fall into this category.

Effects

Looking at the effects of the shutdown will indicate how shutting down the particular equipment affects its surroundings and people who work on or near it. For example, will disconnecting the circuit feeding the lights in a store shut down the illumination to a part of the premises which will cause lost business or be a safety hazard? Will shutting down a piece of machinery affect another process in an assembly line, or the employees that work on the assembly line? This is when a person should ascertain the best time to perform the shutdown. It may need to be performed when the employees who work on or near the equipment aren't around, or when the business is closed. Performing a shutdown when a business is closed will typically increase the cost of the job, as the work will have to be done outside of normal working hours, but it may be the safest and most efficient way to complete the task.

Written Plan

Some LOTO scenarios will require a written plan which details the process that will be followed and the sequence of the work. If this is the case, it will be detailed in the company's energy control program.

Notification of Personnel

Any person who is involved with the piece of equipment or machinery that is to be disconnected must be notified. This notification includes all authorized employees as well as affected employees. Authorized and affected employees are terms which are defined by OSHA: an authorized employee is the person who will install the LOTO device and service a piece of equipment.; an affected employee is the person whose job requires them to work on or around the piece of equipment that is being serviced. They both need to be notified when a shutdown is to occur. It is important to remember that there are often several affected employees, including the equipment operator(s) as well as the people who work around the equipment. There may also be several authorized employees, as more than one person may be servicing the equipment. All people involved with the equipment must know what is taking place as well as how long the work will take to ensure that no worker attempts to perform a startup or release of energy while others are still working on the equipment.

Shut Down Equipment

Some pieces of equipment, such as lights, can simply be turned off, but that isn't always the case. Many large pieces of equipment will have a shutdown procedure which must be followed before the power can be removed from the circuit. Although control devices such as pushbuttons (Figure 6-3) are used to shut down a process, they do not remove power from all parts of the equipment. The isolation of power must be performed by a means of disconnection that completely removes the electric power supply from the equipment, not just the control circuit. (Figure 6-4)

Remove All Energy Sources

Removal of the energy sources includes electrical as well as any other type of source which may be present. Types of energy sources include: electrical, thermal, steam, water, pneumatic, hydraulic, gas, kinetic (spinning wheels or moving objects), potential (spring tension or suspended weight), etc.

Figure 6-3 Pushbutton station used to start and stop a motor.
Courtesy Delmar/Cengage Learning

Figure 6-4 A disconnect switch is used to remove the electric power supply from equipment. *Courtesy Delmar/Cengage Learning*

Electrical Source

The main electrical power may not be the only electrical source for a piece of equipment. Equipment may contain capacitors, batteries, solar panels, or a generator. All of these items must also be disconnected or discharged to completely isolate the equipment from an electrical hazard.

It is important to remember that by simply turning off a disconnect switch an arcing incident can be created, particularly if the disconnect is operated while under load. Always wear the appropriate personal protective equipment, stand off to the side, and follow any other prescribed procedures when operating a disconnect switch.

Non-Electrical Energy Sources

There are many dangers associated with working on machinery and equipment that aren't electrical. There are often several energy sources which a person unfamiliar with the equipment may not be aware of. Mechanical blocks are often necessary to prevent suspended weight from falling or hydraulic pressure being released. Gas, water, air, and other energy sources will need to be removed. (Figure 6-5) Equipment

Figure 6-5 Valves are used to disconnect supplies of gas, water, air, etc.
Courtesy Delmar/Cengage Learning

and machinery with multiple energy sources should have procedures in place for safely removing all of the energy sources, which must be carefully followed to ensure the equipment is in a safe working condition.

Apply Locks and Tags

Once the equipment has been shut down and isolated from all energy sources, the equipment must be locked off to prevent a person from restoring the power while the equipment is being worked on. It is required to not only lock the equipment off using a keyed or combination lock, but to also tag it to indicate that the energy source must remain off. (Figure 6-6)

Lockout

A lockout device is intended to provide a physical restraint which will prevent equipment from being turned on. The lock used can be either a keyed or combination lock. It must only be used for locking off equipment and must not share a key or combination with any other locks.

Figure 6-6 A circuit breaker in this panel has been locked out and tagged out. *Courtesy Delmar/Cengage Learning*

If using a keyed lock, the only person who should have the key is the person performing the task, and likewise the only person who should know the combination to a combination lock is the person performing the task. The lock shall have a method of identifying the worker who installed it.

If there are multiple people working on the same piece of equipment, each person shall install their own lock. A hasp, as shown in Figure 6-7, will allow multiple locks to be installed. The only person who is permitted to remove a lock is the person who installed it.

Tagout

In addition to being locked out, equipment must also be tagged. (Figure 6-8) The tag will identify that the equipment has been locked out for servicing, and should include wording that clearly indicates that there is a danger if the disconnected item is turned back on, and that the only person who should remove the tag is the person who installed it.

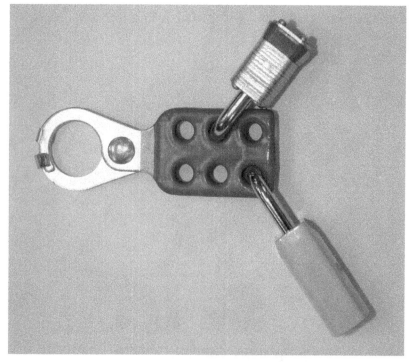

Figure 6-7 Hasp with two locks installed. *Courtesy Delmar/Cengage Learning*

Figure 6-8 Lockout tag. *Courtesy Delmar/Cengage Learning*

It should also include the name of the person who installed the tag and the time and date that the tag was installed. Some tags will include additional information, such as the department that is performing the work, the time the work is expected to be completed, and so on.

The tag should be able to withstand 50 lbs of force at a right angle without breaking, to prevent accidental removal. It shall be attachable by hand, non-reusable, non-releasable, and able to withstand environmental conditions without smearing or fading. If there are multiple people working on the same piece of equipment, each person shall install their own tag.

Verify Equipment Isolation

After the equipment has been locked out and tagged, equipment isolation must be verified. Verifying the absence of voltage can be performed using an electrical meter. (Figure 6-9) Caution must be taken to always check the meter on a known voltage source to verify that it is working properly before and after the test. Non-contact voltage detectors are a handy tool which can be very helpful; however, a voltmeter or voltage tester is the safest way to verify the absence of voltage. Under certain conditions, voltage detectors have been known to give false readings. (Figure 6-10)

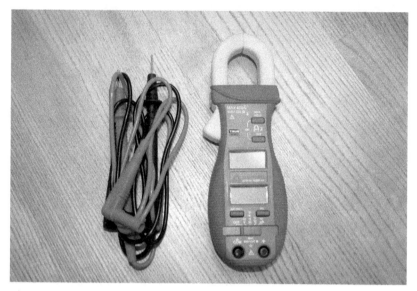

Figure 6-9 A voltmeter can be used to verify the absence of voltage. *Courtesy Delmar/Cengage Learning*

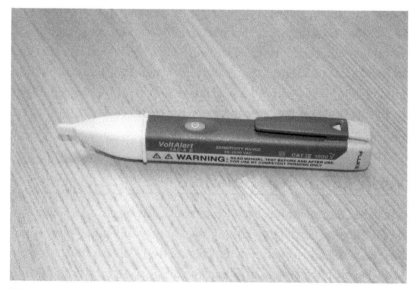

Figure 6-10 Non-contact voltage detector. *Courtesy Delmar/Cengage Learning*

If there are starting controls on the equipment, a person should attempt to start it to confirm that the power is actually disconnected. Push the start buttons, pedals, etc. to verify that the equipment won't turn on.

Apply Safety Grounds (if applicable)

If there is a chance of induced voltages or a backfeed, it may be necessary to install safety grounds. Figure 6-11 shows a safety ground set. The safety ground set is attached to each of the conductors and connected to ground so that if the conductors were to be energized they would immediately short to ground rather than creating a shock hazard. Safety grounds must be sized to carry the maximum fault current for the time necessary to clear the fault. Safety grounds are not used in every scenario, only when warranted.

Restoring Equipment

When the work is complete and the power is going to be restored, there are several steps that must be completed before turning the power back on. This process will be prescribed in the LOTO procedure

Figure 6-11 Ground set. *Courtesy of Salisbury by Honeywell*

which is a part of the employer's safety plan. The following are items which should be part of the procedure.

- Notification of all persons working on or associated with the machine;
- Visual inspection;
- Removal of safety grounds;
- Removal of locks and tags;
- Following the startup procedure.

Notification of Personnel

The first step to be taken once work has concluded and the equipment is ready to be energized is to notify all personnel that could be working on or near the equipment that it is going to be turned on. This means walking around and finding all the authorized and affected employees and ensuring that everyone is clear and ready to go. If the machine has an operator he or she should be tasked with performing any safety checks required before starting.

Visual Inspection

A visual inspection must be performed. This should look for incomplete items such as tools left in, on, or around the equipment or covers left open. At this point it can be verified that every person in the area is clear of the equipment.

Removal of Safety Grounds (if applicable)

If a safety ground set was used it would be removed at this point. If this important step was missed, there will be a short circuit when the power is turned back on, which may have the potential to create an arcing incident.

Removal of Locks and Tags

After all personnel have been notified, the equipment worked on has been visually inspected, and any safety grounds have been removed, it is time to remove the locks and tags. Remember that the only person who can remove a lock is the person who installed it and was performing the work.

Abandoned Lock Procedure

In the event of an abandoned lock on the hasp, there must be a procedure in place to address the situation. This may happen if there was a shift change, or if someone left ill. The procedure must ensure that the absent person's work was completed and that there is no potential danger from the work that person was performing. The supervisor for that particular individual must be able to verify all of the safety procedures before the lock and tag are removed.

Energizing the Equipment

Proper startup procedures must be followed when starting up equipment. This may require notification of surrounding employees, a specific sequence of operation, or the simple flipping of a switch: it all depends on the situation and the equipment. Remember to wear the appropriate personal protective equipment before operating any switches or overcurrent devices.

Lockout-Tagout Devices

There are many different types of LOTO devices on the market. Figures 6-12 through 6-19 show a few commonly-used ones. Almost every scenario imaginable has a lockout device specifically designed for it. The important thing is to use them!

Figure 6-12 Lockout padlocks shall not be used for any other purpose.
Courtesy Delmar/Cengage Learning

Figure 6-13 Lockout tag. *Courtesy Delmar/Cengage Learning*

Figure 6-14 A lockout device to prevent a circuit breaker from being turned back on. *Courtesy Delmar/Cengage Learning*

Figure 6-15 A lockout device used to prevent a switch from being turned on. *Courtesy Delmar/Cengage Learning*

Figure 6-16 A lockout device to prevent a cord from being plugged in. *Courtesy Delmar/Cengage Learning*

Figure 6-17 Lockout hasp. *Courtesy Delmar/Cengage Learning*

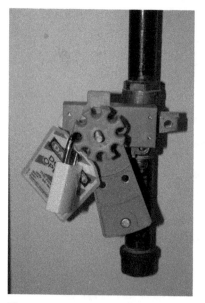

Figure 6-18 A lockout device used to prevent a valve from being turned on. *Courtesy Delmar/ Cengage Learning*

Figure 6-19 A lockout device used to prevent a valve from being turned on. *Courtesy Delmar/ Cengage Learning*

Summary

- Lockout-tagout procedures are in place to prevent unintentional startup or release of stored energy while equipment is being worked on.
- A company's safety plan will have a written procedure for LOTO.
- Planning for a shutdown includes investigating the available energy supplies and looking at the effects of a shutdown, and may include a written plan.
- Before shutting down equipment, all affected and authorized persons must be notified.
- Proper shutdown procedures must be followed before removing power.
- When disconnecting equipment, all energy sources must be removed.

- Equipment must be locked and tagged to prevent energy from being released and to identify who installed the lock.
- Before working on equipment, it must be verified that all energy sources have been removed.

Review Questions

1. What is the purpose of LOTO?
2. List six items that should be included in a LOTO procedure.
3. Who should be notified before shutting down equipment?
4. A tag should be able to withstand _____ lbs of force without breaking.
5. List five items that should be part of a procedure to restore power.
6. Who is permitted to remove a lockout lock?
7. What info should a lockout tag include?
8. How can a person verify isolation of energy sources?

7

Personal Protective Equipment

Objectives

- Explain the need for personal protective equipment;

- Describe the various types of:
  - Eye protection
  - Hearing protection
  - Foot protection
  - Respiratory protection
  - Fall protection

- Explain the proper use of:
  - Eye protection
  - Hearing protection
  - Foot protection
  - Respiratory protection
  - Fall protection

Introduction

Personal protective equipment (PPE) is an extremely important part of everyday life for electricians and anyone else working on a construction site. There are many dangers that can be avoided by simply wearing the appropriate personal protective equipment. The level of PPE that must be worn will vary according to the working conditions and the task being performed.

This chapter is dedicated to PPE which is general in nature, and used by all workers on a construction site. Personal protective equipment specifically for electrical hazards is covered in Chapter 5.

Before we look at the various types of PPE, there are a few important points that must be mentioned. First, in order for PPE to offer protection, it must be used. This may sound obvious, but it is quite common for workers to bring PPE with them but not put it on. For example, a worker who is simply walking through an area or whose task doesn't have the potential to create flying objects may not wear his safety glasses, instead keeping them on top of his head or in his pocket, but a person working nearby could hit a nail with a saw and cause a piece of metal to fly into the first worker's eye. It is important to remember that you aren't the only person working on the jobsite. The hazards could just as easily be caused by someone or something else.

The proper use of personal protective equipment is also essential. If PPE isn't used properly it may not offer the necessary protection, exposing the worker to dangers. For example, if a harness isn't put on or tied in properly, it won't offer fall protection. The employer has the responsibility to train employees in the proper use of PPE. If a worker hasn't used a specific type of PPE for an extended period of time, retraining will be required.

Personal protective equipment must be inspected regularly. Like anything else, PPE will wear out with use. If wear or damage that could jeopardize the integrity of the equipment is found, the gear must be removed from service for repair or be replaced immediately.

Eye and Face Protection

In the United States, approximately 2,000 job-related eye injuries occur every day. The majority of these injuries could be avoided with proper eye protection. OSHA requires eye protection when there is a reasonable chance that an eye injury could occur without it. People working in the construction industry are at risk for eye injuries on a daily basis because of the type of work they are doing, or is being performed around them.

There are many different types of protection for the eyes. Safety glasses, safety goggles, and face shields are the three that will be discussed in this text.

Safety Glasses

Safety glasses are intended to prevent flying debris from hitting the eye. (Figure 7-1) Electricians should wear safety glasses at all times, as nearly everything they do on the job has the potential to create flying debris.

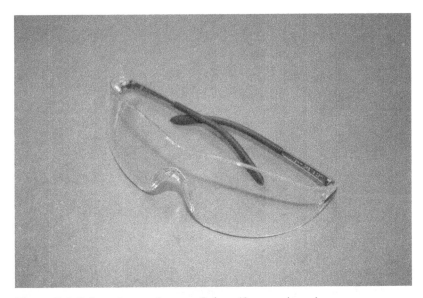

Figure 7-1 Safety glasses. *Courtesy Delmar/Cengage Learning*

Hammering, drilling, sawing, cutting, etc. all create flying debris which could cause an eye injury. (Figure 7-2) Don't forget the danger of arcing incidents creating flying particles. All of these situations will happen with no warning at all.

Many workers will have safety glasses with them but only put them on when they are performing a task that could produce flying debris. This is a dangerous habit to fall into, as your particular task is often not the only hazard present. The person working near you may be drilling or hammering, or might simply drop something on the floor, causing a particle to become airborne. The inconvenience of wearing safety glasses is minor compared to losing one's eyesight.

Choose safety glasses that are comfortable. One of the reasons that people often don't wear them is that they find them uncomfortable. By taking the time to try on several pairs and find the ones that fit the best, you will find that you won't be bothered by wearing them.

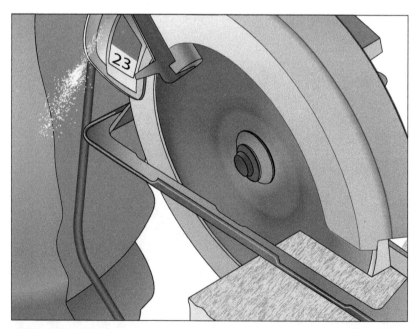

Figure 7-2 Saws and other power tools create flying debris. *Courtesy Delmar/ Cengage Learning*

It is important to create good habits. By wearing them all the time, and not just when there is a danger, they will start to become part of your everyday work attire. You won't mind wearing them; in fact, without them you will feel as if something is missing. It is also important to lead by example. As others see you wearing glasses they will also tend to start wearing them.

What to Look for in Glasses

Safety glasses must be designed and listed as safety glasses, with shatter- and impact-resistant lenses. Safety glasses should have side shields or curve around the side of the face to protect the eye from objects that may come flying from the side.

Prescription glasses and sunglasses which are not listed as safety glasses will not meet the shatter and impact requirements of safety glasses; if they are hit by a flying object it may go through or cause the lens to shatter. Prescription safety glasses are available, that often have detachable side shields which can be removed when not at work.

Safety glasses are available with anti-fog, anti-scratch, and anti-static coatings. All of these are great features that will make wearing the glasses more comfortable. Wearing safety glasses that are fogged or scratched will create a safety hazard. They obstruct a person's view, which can lead to an accident, and a person will be more likely to not wear them, which could lead to an eye injury.

It is important to take good care of safety glasses. Throwing them in with tools or laying them lens down on the floor will quickly lead to scratched lenses which are hard to see through; this in itself creates a safety hazard. Storing the glasses in a pouch or case in a safe place will help to keep them in good condition.

Safety glasses are also available with tinted lenses and UV protection, great features when working outdoors. Keep in mind that sunglasses are not the same as safety glasses, as they do not have the correct lenses and frames designed to protect the eyes from flying objects. It is also important to be aware that wearing tinted safety glasses indoors can be dangerous, as the tint may impair a person's vision, causing a safety hazard.

Safety Goggles

Safety goggles can provide additional protection over safety glasses, as they can guard against such dangers as splashed liquids, airborne dust, particles falling from above, etc. (Figure 7-3) Falling debris is a common occurrence when working on a construction site. It may come from you or someone near you drilling overhead, people above you sweeping a platform, and so on. Regular safety glasses do not offer protection against particles falling from above; they may simply fall between your glasses and your face, and can get into your eye, potentially leading to a scratched cornea. People who wear prescription glasses which are not listed safety glasses may wear safety goggles over their glasses to provide the necessary protection.

Safety goggles are available in several different styles to meet various situations. Goggles are available with prescription lenses for workers who don't want to wear their goggles over a pair of prescription glasses. Like safety glasses, safety goggles can be purchased with anti-fog, anti-scratch, and anti-static coatings.

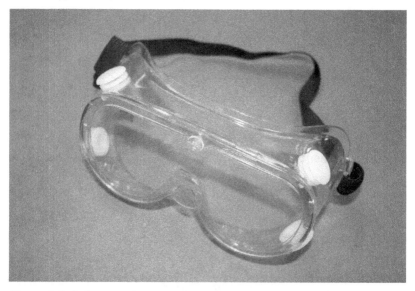

Figure 7-3 Safety goggles. *Courtesy Delmar/Cengage Learning*

Face Shield

When performing certain tasks, safety glasses or goggles alone may not provide the necessary protection for a person's face and eyes. (Figure 7-4) Examples include cutting metal with a chop saw or using a grinder: there will be hot metal particles flying around which can hit a person's face and may even ricochet into their eye. (Figure 7-5) Workers are still required to wear their safety glasses under the face shield for added protection. Like with safety glasses and goggles, it is important to take good care of face shields to prevent them from becoming scratched and difficult to see through. Impaired vision from a scratched face shield is a recipe for disaster while working with power tools.

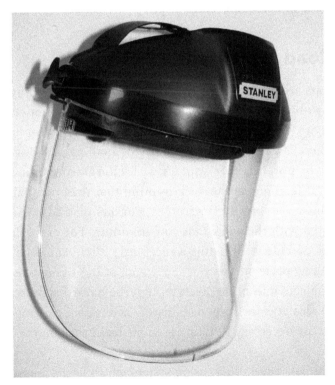

Figure 7-4 Face shield. *Courtesy Delmar/Cengage Learning*

Figure 7-5 A face shield is necessary when using power tools such as chop saws. *Courtesy Delmar/Cengage Learning*

Head protection

Hard hats are designed to prevent head injuries caused by falling or flying objects, or from bumping into hard surfaces or energized conductors. (Figure 7-6) OSHA requires workers to wear hard hats when there is a reasonable possibility of a falling or flying objects hazard, or potential for contact with energized conductors or objects. It is important to keep in mind that the potential hazard from above is typically from other workers or trades that aren't associated with the work that you are doing. For example, while working outside terminating a receptacle, there could be shingle work taking place on the roof above you, or there could be a crane lifting objects into place. Even though the hazard is presented by other trades, it is the responsibility of each individual employer to provide the necessary PPE for its employees based on the hazards present.

Hard hats have a suspension system which keeps the hard shell at least 1¼ inches away from the workers head. (Figure 7-7) This

Figure 7-6 Hard hat. *Courtesy of MSA*

space is necessary for the hard hat to be able to absorb an impact from above without causing injury. Inserting gloves or other items into this space can be dangerous, as an impact will be transferred directly to the worker's head.

Type and Class

Hard hats come in two types, Type 1 and Type 2. Prior to 1997, the two types designated the brim style: Type 1 meant the brim encircled the helmet, while Type 2 had the brim only on one side or none at all. In 1997 the "Type" classification changed: it no longer refers to the brim style, but now denotes the type of impact protection. Type 1 helmets offer protection from blows coming from the top, while Type 2 helmets offer protection against blows from the top or side.

Hard hats are available in three classes: prior to 1997, the classification for hard hats was Class A, B, and C. Class A hard hats offered impact protection as well as voltage protection to 2,200 volts.

Figure 7-7 Hard hat suspension system. *Courtesy Delmar/Cengage Learning*

Class B hard hats offered impact protection as well as voltage protection to 20,000 volts. Class C hard hats offered impact protection without any electrical protection. In 1997 the classification of hard hats was changed, and they are now classified as Class G, Class E, and Class C.

- **Class G** hard hats are designated as general hard hats. They are the same as the old Class A and provide impact protection as well as electrical protection tested to 2,200 volts.
- **Class E** hard hats are designated as electrical hard hats. They are the same as the old Class B and provide impact protection as well as electrical protection tested to 20,000 volts.
- **Class C** hard hats are designated as conductive hard hats. They are the same as the old Class C and provide impact protection without any electrical protection.

Most hard hats are intended to be worn with the brim facing forward, but there are some which are rated to be worn with the brim facing forward or backward. These helmets will carry a reverse donning symbol inside the helmet. (Figure 7-8)

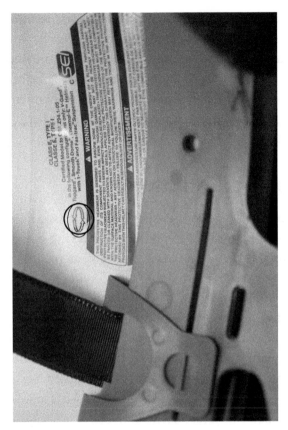

Figure 7-8 This helmet has the reverse donning symbol and has been listed to be worn with the brim forward or backward. *Courtesy Delmar/ Cengage Learning*

Hard hats which are to be worn in cold temperatures (−30 degrees Fahrenheit and below) should have a low temperature rating, and will be marked "LT" inside the helmet. If a regular helmet is worn in these extreme temperatures, there is a chance of it shattering upon impact and not providing the intended protection.

Ultraviolet radiation from the sun or other sources is damaging to a hard hat over time. Damage from UV will appear as a dull or chalky surface on the helmet. When not in use, avoid storing the hard hat in direct sunlight. Workers who spend a lot of time in the sun will have to replace their hard hats more frequently.

Hard hats should be inspected daily for cracks, penetrations, dents, or any other damage which could be due to impact, wear, or abuse. The suspension should also be inspected daily for fraying, tears, and any other signs of damage. If any damage is found, the helmet should be discarded immediately. Some guidelines recommend replacing a helmet's suspension once a year, and replacing a hard hat after five years of use.

Hearing Protection

Work in the construction industry is noisy: jobsite noise levels are often high enough to cause permanent hearing damage. A good rule of thumb is that if you have to raise your voice to have a conversation with someone standing next to you, the noise levels are dangerous.

The unfortunate truth about hearing damage is that it is irreversible: once the damage has been done, there is no way to correct the problem. Hearing damage is not something that shows up immediately: it may take years before hearing loss is noticed, and by then it is too late to do anything about it. The only way to protect yourself is to wear the appropriate protection whenever it is necessary.

Noise levels are expressed in decibels, and anything above 85 decibels can cause hearing damage. A ten-decibel increase represents a doubling of the noise level received by the human ear. OSHA has created guidelines which indicate the amount of time a person can be safely subjected to a given noise level. (Figure7-9) Most of the tools used in the construction industry exceed the levels that are safe for the human ear. (Figure 7-10)

Ear protection is intended to reduce exposure to noise. Each protective device will have an NRR (Noise Reduction Rating). The NRR indicates the number of decibels the noise level will be reduced through use of the ear protection. In some extreme situations, more than one type of ear protection must be used together to bring the noise down to a safe level. There are some ear protection devices on the market that will allow low noise levels, such

OSHA Permissible Noise Exposures	
Sound Level	**Duration Per Day**
90 dB	8 hours
91.5 dB	6 hours
93 dB	4 hours
94.5 dB	3 hours
96 dB	2 hours
97.5 dB	1 1/2 hours
99 dB	1 hour
102 dB	1/2 hour
105 dB	1/4 hour or less
105+ dB	extreme risk

Figure 7-9 OSHA Permissible Noise Levels. *Courtesy Delmar/Cengage Learning*

as conversations, to filter through but will block high noise levels. There are two main types of ear protection: ear plugs and ear muffs.

Ear plugs fit into the ear canal to block the noise. (Figure 7-11) They may be reusable or disposable. One common type is foam ear plugs, which are simply squished up and inserted in the ear canal. After they are inserted they will start to expand back to their original size, sealing the ear canal. This type of protection

Noise Level of Common Tools	
Tool	**Noise Level**
Corded Drill	98 db
Circular Saw	110 db
Hammerdrill	114 db
Grinder	98 db
Air Compressor	90 db

Figure 7-10 Noise levels of common tools used on a construction site. *Courtesy Delmar/Cengage Learning*

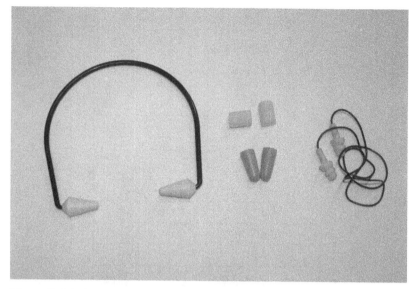

Figure 7-11 Ear plugs are available in various styles. *Courtesy Delmar/ Cengage Learning*

is disposable, and may irritate the ear canal when used for long periods of time.

Custom-molded ear plugs are made by a professional who fits the plugs perfectly to an individual's ear canal. This type of protection is more costly, and care must be taken to properly clean them and keep them in a safe place when not in use.

Ear muff-type protectors sit on the outside of the ears, held by a strap which goes over the top of a person's head. (Figure 7-12) Since the strap goes over the head, regular ear muff protectors will not work with a hard hat. To solve this problem, most hard hats have provisions for adding ear muff-type protectors to the hard hat itself. Ear muffs which are designed to be used with a hard hat will simply clip into a groove on the helmet. (Figure 7-13)

Ear muff protection can be used in conjunction with ear plugs to give added protection. This will often be necessary when working in an area where the noise levels are extreme and one type of protection will simply not bring the noise down to a safe level.

Figure 7-12 Ear muff type hearing protection. *Courtesy Delmar/Cengage Learning*

Figure 7-13 Hard hat with attached hearing protection. *Courtesy of MSA*

Work Boots

Construction workers find themselves working on a variety of surfaces, such as uneven or soft soil, sticky mud, uneven boards, corrugated tin, concrete, etc. Uneven or loose surfaces can often lead to ankle injuries, while working on concrete or other hard surfaces can cause foot or leg fatigue and lower back pain. There is also the real possibility of heavy objects falling on construction workers' feet. Work boots are intended to protect a person from these hazards. (Figure 7-14)

There are several types of work boots on the market, all designed to meet various needs. It may be necessary to have more than one type of work boot so that the appropriate footwear can be worn in different environments. For example, when working outside in the cold of winter, a worker may want an insulated boot, but when working indoors or in the summer heat an uninsulated boot will be more comfortable.

Good-quality work boots which are properly fitted will provide protection for your foot as well as your lower back. Standing on hard surfaces for extended periods of time can cause a person's feet to hurt,

Figure 7-14 Workboots. *Courtesy Delmar/Cengage Learning*

legs to become tired, and lower back to ache. Proper work boots will offer the support needed to help avoid some of these discomforts.

Cheap, poorly-made footwear will not fit properly, not allow proper blood flow, and cause blisters, among other issues. Spend the money necessary to get good footwear that will offer the necessary protection and comfort. Your feet and back will thank you every day.

It is recommended to have two pairs of boots which can be alternated from day-to-day. This allows enough time for the boots to dry out and allows the sole to decompress. Although the initial investment of having two pairs of boots may be a large one, it is better for your feet, legs, and back. It is also better for the boot and will extend its life. It is best to get fitted for a new pair of boots at the end of a workday, when your feet will be tired, sensitive, swollen and more in tune with the proper fit.

Ankle Protection

Some boots offer ankle protection, with taller sides that will help prevent a worker from rolling an ankle when stepping on objects or loose soil. Construction sites are dirty and cluttered, making it very easy to roll an ankle: these work boots will offer a level of protection against this.

Safety Toe

With the danger of falling objects on a jobsite, a safety toe is a must. In fact, many sites require that all workers have footwear with toe protection. In the past, steel toes were the most common type, but there are now other materials available which will provide the necessary toe protection.

A common misconception is that a steel-toe boot will be uncomfortable, with the toe rubbing on an edge in the boot, but a worker wearing a properly-sized, quality work boot won't even notice that they are wearing toe protection.

EH Rating

Work boots are available with an electrical hazard (EH) rating. These boots have an insulated sole intended to protect a worker against current flow which could pass through the boot into the ground. Unfortunately, the only time this protection is ensured is when the boots are first taken out of the box. After walking around a construction site strewn with pieces of metal, nails, etc., there is a likelihood that the insulating qualities will be degraded.

However, even with the possibility of a nail or other debris jeopardizing the insulation of EH-rated boots, they are still a must for electricians. The additional protection may be just what it takes to save your life.

Work Gloves

Work gloves come in many styles and types to meet workers' various needs. (Figure 7-15) Jobsites are often dirty, oily, and full of sharp and pointed objects; it is important to have work gloves that are suited to the job being performed and are sized properly. Most general

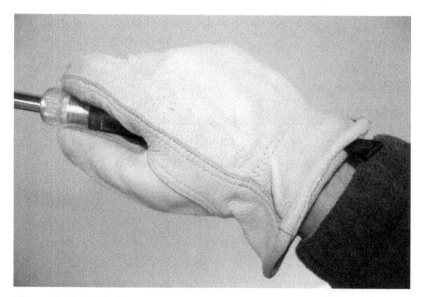

Figure 7-15 Work gloves. *Courtesy Delmar/Cengage Learning*

construction work can be completed with many types of gloves, in which case it is a matter of personal preference and proper fit.

Respiratory Protection

When working in dusty environments, or when there are airborne particulates present, it is necessary to wear a mask or respirator. Construction sites are often dusty places, particularly when remodeling or crawling around in an insulated attic.

Mask

It is very common to have a package of masks in a work truck or gang box. These masks have a metal tab which can be formed to the nose to provide a good seal. (Figure 7-16) Some masks are simply made from a filtering material through which a person inhales and exhales, while others have a one-way valve which makes exhalation easier and doesn't retain as much of the heat and moisture from a person's breath. (Figure 7-17) Regardless of the type of mask, it is important to ensure a good seal or the mask won't do much good.

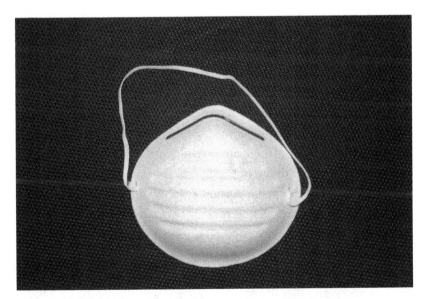

Figure 7-16 Particulate mask. *Courtesy Delmar/Cengage Learning*

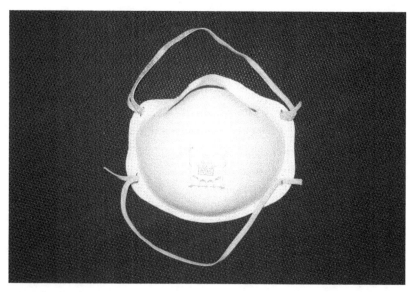

Figure 7-17 Particulate mask with one way valve. *Courtesy Delmar/Cengage Learning*

Although dust masks work well for many types of particles, they do not provide the same protection as a respirator.

Respirator

The type of respirator most commonly used by construction personnel is the type which simply filters the air. (Figure 7-18) This type of respirator consists of a rubbery seal which is pressed against the face, a one-way exhalation valve, and filtering cartridges. Cartridge-type respirators are designed to prevent the user from inhaling harmful dust, vapors, fumes, or gases. There are several types of filters available, each designed to filter out different particles, vapors, etc. It is important to pick the correct filter for the particular job that is being done. Filters do not last forever: be sure to replace them at the intervals recommended by the manufacturer based on the type of environment they are exposed to.

Before each use, it is important to check for a proper seal around the face and verify that the one-way valve is working properly. Always

Figure 7-18 Half face respirator. *Courtesy Delmar/Cengage Learning*

follow the manufacturer's recommendations for the proper testing, use, and care of a respirator.

There are respirators available that supply fresh air to the worker through a hose which attaches to the respirator. This may be necessary in some situations where oxygen levels are low or there is some other air quality issue that won't be addressed by simply filtering the air.

Fall Protection

Many times construction workers will find themselves working from elevated platforms or on ladders, scaffolds, etc. Falls are the leading cause of death in the construction industry, accounting for one-third of all fatalities. OSHA dictates that if a person is working six feet above a lower level, or runs the risk of falling into dangerous equipment, they must have fall protection in place.

The protection may consist of guardrails, safety nets, personal fall arrest systems, or other measures. As this chapter is about personal protection equipment, it will focus on personal fall arrest systems.

Personal Fall Arrest Systems

A personal fall arrest system consists of the anchor point, a lanyard or rope, a harness, and the connecting hardware. Personal fall arrest systems are very common in the construction industry, as workers are often working from elevated locations. Workers must be trained on the proper use of fall protection products before attempting to employ them.

A personal fall arrest system must not allow a person to free-fall more than six feet before it begins to stop the fall. If a platform or piece of equipment is less than six feet below, then it must stop a person before they make contact. Once the system begins to stop a person from falling, it must slow the person down in a distance of 3½ feet or less.

Equipment associated with a personal fall arrest system must be inspected for wear, deterioration, and damage prior to each use, and in accordance with the manufacturer's recommendations. If any damage or wear is found, the system should be removed from service immediately. Any equipment which has been involved in a fall must also be removed from service immediately. Some equipment has built-in indicators which show if it has been involved in such an incident.

Anchor Point

The anchor point is a secure point to which the lanyard or lifeline is tied. (Figure 7-19) It must be capable of supporting 5,000 pounds per person attached to it, or be a part of an engineered fall arrest system. Care must be taken to use the appropriate connector to fasten to the anchor point and ensure that there aren't any sharp edges which could cut through the rope or lanyard in the case of a fall. There are many anchor connectors on the market to meet various situations. The location of the anchor point should be above the worker to avoid a swing fall.

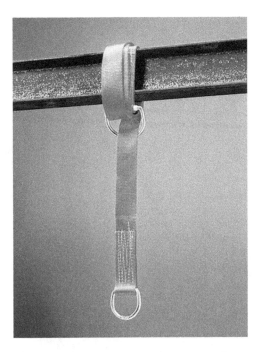

Figure 7-19 Temporary anchor connector.
Photo courtesy of Miller® Fall Protection

Lifelines

Lifelines are a flexible line running either horizontally or vertically. A horizontal lifeline will have an anchor point at both ends of the line. (Figure 7-20) The worker's lanyard will be attached to the lifeline with the appropriate hardware. This system offers great mobility, as the worker's lanyard can slide along the length of the lifeline. This type of system may be used when there are no anchorage points above the worker. Vertical lifelines have one anchor point at the top of the line. (Figure 7-21) Vertical lifelines are often used with ladders and other situations where a worker will be ascending and descending. The worker's lanyard is connected to the vertical lifeline with a rope or cable grab which will slide up and down the lifeline as the worker moves. (Figure 7-22) In the case of a fall, the grab is designed to bind and stop the worker from falling. Only one person can be attached to a vertical lifeline at any time.

Figure 7-20 Horizontal Lifeline. *Photo courtesy of Miller® Fall Protection*

Self-Retracting Lifeline

A self-retracting lifeline will lengthen and retract with a worker's movements, while still arresting the worker in the event of a fall. (Figure 7-23) This allows the worker more mobility. Self-retracting lifelines contain a braking system which will decelerate and stop a person within a couple of feet, and are available with cables or straps to best suit a particular task.

Lanyard

A lanyard is a rope, cable, or strap which connects a worker's safety harness to an anchor point or lifeline. (Figure 7-24) Shock-absorbing lanyards have an incorporated deceleration device which will slow the worker's fall and absorb some of the energy. It is important to keep in mind that a

Figure 7-21 Vertical Lifeline. *Photo courtesy of Miller® Fall Protection*

Figure 7-22 Robe Grab. *Photo courtesy of Miller® Fall Protection*

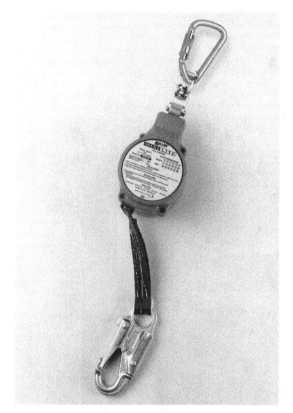

Figure 7-23 Self Retracting Lifeline. *Photo courtesy of Miller® Fall Protection*

shock-absorbing lanyard will grow in length by as much as 30 inches as it absorbs a fall. When determining the type of fall protection needed and the distance a person might fall, this must be kept in mind.

Harness

A body harness is a configuration of straps which will stop and support a person in the case of a fall. The design of body harnesses will distribute the forces present during the deceleration of a fall so they are not concentrated on one part of the body. (Figure 7-25) The back of the harness will have a D-ring where the lanyard is connected to the harness. If the harness has D-rings on the front, they are for

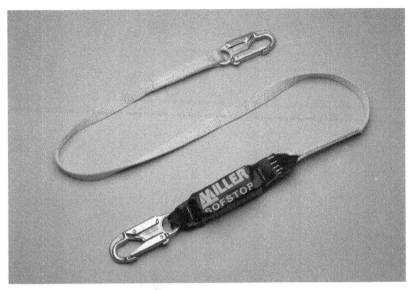

Figure 7-24 Shock Absorbing Lanyard. *Photo courtesy of Miller® Fall Protection*

positioning and retrieval only. It is important to properly don a harness; all straps must be in the right position, and be snug without limiting range of motion.

Body belts are no longer permitted for fall protection, as dangerous forces can be exerted on a person's midsection, causing internal damage. There is also the potential for a person to flip over.

Safety Nets

Safety nets are designed to catch a person and stop their fall before they reach the ground or another surface. Safety nets should be installed as close as possible to the height of the worker, but must not be more than 30 feet below him or her. The potential fall area must be unobstructed, to prevent worker injury during the fall. The openings in the net must not exceed 36 inches square or 6 inches on any side, and all mesh crossings must be secured to prevent the openings from enlarging.

Safety nets must be drop-tested to ensure that they will offer the necessary protection. This is done by dropping a 400 lb bag of sand, roughly

Figure 7-25 Harness. *Photo courtesy of Miller® Fall Protection*

30 inches in diameter, into the net. This must be performed after the initial installation but prior to use, after it has been relocated, after a repair, and every six months if it is up for prolonged periods of time.

Safety nets must be inspected for damage at least once a week. If any damage is found, the net must be removed from service immediately. If any tools or materials fall into the net, they must be removed immediately, as they could injure a worker or sever the net in the event of a fall.

Summary

- There are many dangers on a construction site that can be avoided by wearing the appropriate personal protective equipment.

- Workers must be trained on the proper use of personal protective equipment before attempting to use it.
- Eye protection is necessary in almost all situations on a construction site.
- Where there is a possibility of falling or flying objects a hard hat must be worn.
- Most tools and equipment on a construction site produce dangerous noise levels. Hearing protection must be worn to prevent permanent hearing damage.
- Quality work boots will protect workers' feet, legs, and back from the stress of standing on hard surfaces all day and from hazards underfoot.
- Certain types of work on construction sites will produce airborne particles and will necessitate the use of respiratory protection.
- A lot of work on a construction site is done from elevated locations. The appropriate fall protection is necessary to prevent life-threatening falls.

Review Questions

1. List five tasks which could produce flying debris.
2. When should safety goggles be worn?
3. Describe the protection offered by a Type 1, Class E hard hat.
4. What is the rule of thumb for needing hearing protection?
5. What does NRR stand for?
6. List three types of protection that may be offered by work boots.
7. Before using respiratory protection, it is necessary to check for

   _____.

8. Fall protection is necessary when working _____ or higher above a platform or the ground.
9. What is the advantage of using a self-retracting lifeline over a lanyard and anchor point?

CHAPTER 8

Tool and Equipment Safety

Objectives

- Describe the dangers associated with the various tools and equipment used on a construction site;

- Explain the safety precautions necessary when using hand tools;

- Explain the safety precautions necessary when working with power tools;

- Describe the correct use of step and extension ladders;

- Explain the safety precautions necessary when working with aerial lifts;

- Explain the safety precautions necessary when working with construction machinery.

Introduction

Electricians use tools of one kind or another on a daily basis. Tools save a lot of time and make our work much easier. The benefits that come along with using tools don't come without a price, however, as tools can create dangerous situations. When working with a tool it is important to have undergone any necessary training, follow all recommended safety precautions, and give the tool the respect it deserves.

Hand Tools

Hand tools are those which do not use electricity or any form of power, except human muscle, to perform a task. (Figure 8-1) The work accomplished with hand tools typically involves prying, cutting, twisting, or hammering. All of these can lead to flying objects if something breaks or gives way. Safety glasses should be worn at all times when working with hand tools.

The majority of an electrician's work involves hand tools. Good quality professional hand tools will last longer and be safer. Inexpensive or worn tools will cause a person to squeeze or push harder to

Figure 8-1 Hand Tools. *Courtesy Delmar/Cengage Learning*

accomplish a task than they would with a quality tool in good working order. Using excess pressure often leads to slippage, which can frequently result in injury.

It is also important to keep tools sharp. A dull blade on a utility knife may seem less dangerous, but that is not the case: a dull knife requires more pressure to cut, which increases the chances of slippage. If the dull knife slips under the extra force required, the resulting cut will be worse. Although good tools will last a long time, they do not last forever. It is important to replace tools as they wear out and start to slip or not cut properly.

Improper use of hand tools can lead to injuries, tool damage, and equipment damage. Misuse will tend to cause a person to slip which can result in a pinch or cut. In addition to the potential personal injury, misuse of tools will void their warranty.

Remember the dangers of working around electricity. Always use the proper procedures to shut off a circuit before working on it. If working on a live circuit is justified, be sure to wear the appropriate personal protective equipment and use voltage-rated tools designed for the specific task at hand. The rubber or plastic handles found on most tools are designed for comfort, not electrical shock protection. (Figure 8-2)

Power Tools

Power tools use some external form of power, whether from electricity, air, gas, or chemicals (powder actuation). Power tools are extremely helpful to construction workers, as they save a lot of time and manual labor. They do, however, create some safety hazards. Almost all power tools produce flying objects, so safety glasses must be worn at all times. Power tools also create a lot of noise, so hearing protection will likely be required.

Some power tools are heavy. Be sure to follow proper lifting and working techniques to avoid injuries.

- Lift with your legs, not your back;
- Hold the weight close to your body when carrying the tool;

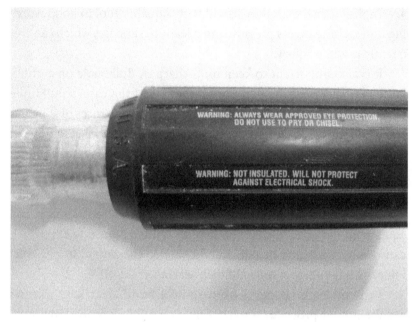

Figure 8-2 The handles on regular hand tools are designed for comfort and do not provide any voltage protection. *Courtesy Delmar/Cengage Learning*

- Do not twist your body while carrying heavy objects;
- Ask for help if the object is too heavy or cumbersome;
- Be sure to secure the tool before operating it.

Just as power tools perform work very quickly, they can also do damage very quickly. Always look for hidden hazards where a power tool is to be used. Just a few hidden hazards include wires in a wall, flammable atmospheres, water lines, gas lines, or nails in a board that is to be cut, but the list could go on. Always be alert and thinking about what potential dangers exist, and how to avoid them.

Electric Tools

Electrically-operated tools can be extremely dangerous. They will typically have sharp components used to cut or drill, and they operate at high speeds. (Figure 8-3) The use of these tools can lead to all sorts of damage and injuries if the necessary precautions and care aren't taken.

Figure 8-3 Reciprocating Saw. *Courtesy Delmar/Cengage Learning*

Be sure to have proper training on each particular tool before using it, know what to expect from the tool, and follow all the manufacturers' recommendations for proper tool usage and PPE.

Before using any electric tool it should be inspected for damage. If any damage is found, it must be marked "do not use" and taken out of service for replacement or repair. Examples of possible damage could be exposed conductors in the cord, a broken blade guard, a cracked handle, and so on.

Electric tools will create flying debris and may produce a lot of noise. (Figure 8-4) Always use safety glasses, ear protection, and any other PPE necessary to perform the task. If there are any other workers nearby that could be affected, inform them of the dangers and the PPE they should employ.

Before plugging in a tool or installing a battery, ensure all switches are in the "off" position. When a particular task has been completed, be sure to unplug the tool to prevent accidental startup. Always unplug or remove the battery before changing bits or blades. Bumping a switch and unintentionally starting up a tool while changing a saw blade or installing a drill bit can cause serious injuries.

Figure 8-4 Hammerdrill drilling into concrete. *Courtesy Delmar/Cengage Learning*

Always use the proper tool for the job. Using the wrong tool can lead to personal injury or property damage. The correct tool will perform the task faster and more safely than attempting it with the wrong tool. Be sure to use safety guards and verify that they are working correctly. Most power tools incorporate safety devices to help prevent injuries; removal of these guards is a recipe for disaster.

Be sure to have good solid footing, balance, and a secure grip on the tool. Many tools will have a side or auxiliary handle to help control them. (Figure 8-5) Using this handle will help ensure that the tool doesn't get away from you and cause an injury. If possible, brace the tool against a fixed object. Overextending to perform a task or failure to maintain a secure footing and grip on the tool may lead to falls or loss of control over the tool. Examples of tools where this is a risk are right-angle drills and hammer drills. Knowing what to expect from the tool is extremely important so that proper precautions can be taken.

Wear the appropriate clothing. Loose clothing, jewelry, gloves, and the like can get tangled up in moving parts. Great care must be taken

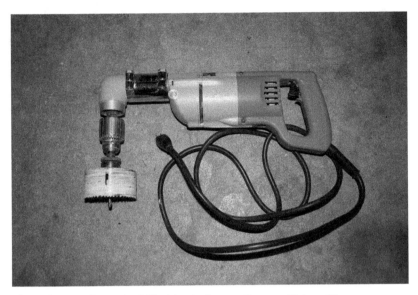

Figure 8-5 Right angle drill with a hole saw. *Courtesy Delmar/Cengage Learning*

to ensure that clothing and other objects are clear from the moving parts of tools such as drills. Drills and similar tools have a lot of torque and run at high speeds. By the time you notice that an article of clothing is caught it will be too late: cuts, broken bones, and other injuries can occur faster than a person can react.

Electric tools should be plugged into a receptacle that has GFCI (Ground Fault Circuit Interrupter) protection. (Figure 8-6) If a GFCI receptacle isn't available, cords with internal Ground Fault Circuit Interrupters are available. (Figure 8-7)

Before use, always inspect the power cords on electric tools for wear or damage. If any damage is found the tool must be removed from service immediately. Never carry or hoist the tool by the cord. If the tool needs to be hoisted to an elevated platform, tie a rope around the handle to raise it. When unplugging a cord from the wall, pull on the plug, not the cord. Pulling on the cord will cause it to separate from the plug, creating a shock hazard. (Figure 8-8) When the work has been completed, unplug the tool and roll up any cords.

Figure 8-6 Tools on a jobsite must be plugged into receptacles that offer GFCI protection. *Courtesy Delmar/Cengage Learning*

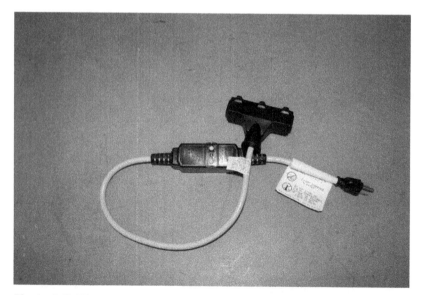

Figure 8-7 Three way splitter that provides GFCI protection. *Courtesy Delmar/Cengage Learning*

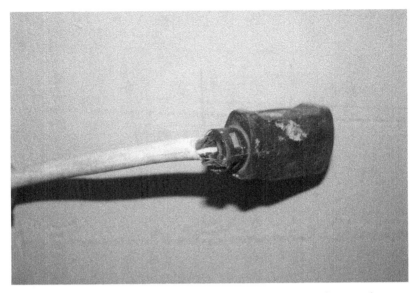

Figure 8-8 This cord has exposed conductors where the cord enters the cap. This is often caused by pulling on the cord rather than the cap when it is being unplugged. *Courtesy Delmar/Cengage Learning*

Cords lying around create a trip hazard, and will get stepped on, run over, and buried under objects, all of which could damage them. (Figure 8-9)

Tools which aren't double-insulated will require a grounding connection to prevent the metal components from becoming energized in the case of a fault. If the ground pin has been removed from or broken off a tool or extension cord, it becomes an electrical shock hazard, and the tool or cord must be removed from service for repair. (Figure 8-10) Double-insulated tools do not require a grounding connection. They will have non-metallic cases, be marked as double-insulated, and have a two-wire cord.

Pneumatic Tools

Although most electricians don't use pneumatic tools on a daily basis, they will often be working near others using them. The pneumatic tool most often used on a construction site is an air nailer. (Figure 8-11)

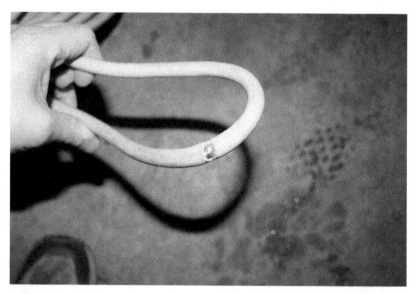

Figure 8-9 This extension cord has exposed conductors due to the cord sheath being cut. Splits in the sheath are often caused by pulling the cord across a sharp object, having it buried under construction materials, or being driven over. *Courtesy Delmar/Cengage Learning*

Figure 8-10 The grounding terminal has been broken off this cord. *Courtesy Delmar/Cengage Learning*

Figure 8-11 Air Nailer *Courtesy. Delmar/Cengage Learning*

Be sure to have proper training on each particular pneumatic tool before using it, know what to expect from the tool, and follow all manufacturer's recommendations for proper tool usage and PPE. Before using any pneumatic tool, it should be inspected for damage. If any is found, it must be marked "do not use" and taken out of service for replacement or repair.

Pneumatic tools can create flying debris as well as loud noises. Always use safety glasses, ear protection, and any other PPE necessary to perform the task. Nail ricochets and pieces thrown from the impact of a nail will create flying debris, so safety glasses are a must. Pneumatic tools use an air compressor to power the tool. These compressors are typically close to the work location and produce a lot of noise. The amount of time a compressor runs depends on the size of the air tank, how often the pneumatic tool is used, and how many air leaks there are. If the compressor runs frequently, hearing protection will likely be necessary. Remember that even if you aren't the one using the tool, flying debris and loud noises can injure you.

Before connecting an air line to a tool, ensure that all the safety guards are in place and the trigger isn't depressed. Never remove or tie back the safety devices, as that could lead to unintentional firing of the tool and serious injury or death. Never carry an air nailer with the trigger depressed. Simply bumping the safety guard into an object could allow the gun to fire. When the job is completed, disconnect the air line from the tool to prevent it from accidentally firing if bumped. Before making any adjustments or removing a jammed nail, disconnect the air hose and ensure that the air has been released. When working around air nailers, be sure not to bump into them with a foot or any other part of the body: under the right circumstances the nailer could fire. (Figure 8-12)

Figure 8-12 Care must be taken when walking around on a job site to avoid kicking or bumping into tools that could cause injury. *Courtesy Delmar/ Cengage Learning*

Never aim air nailers at yourself or someone else; they must be treated with the same respect as a firearm. When using an air nailer, it is extremely important to keep hands and other body parts away from the location being nailed, as the nail could ricochet. Many people have nailed their hand into a board or shot a nail into a part of their body. Be sure to have a good footing, never overextend, and keep a secure grip on the tool to prevent slips, falls, and loss of control.

There are also battery-operated and propane-operated nail guns which require the same respect and precautions as pneumatic nail guns.

Powder-Actuated Tools

Powder-actuated tools are used to fasten materials to concrete or metal. (Figure 8-13) Since they do not require the use of electricity, they are very handy on construction sites and other locations where power isn't available. Powder-actuated tools use a charge, very similar to that of a firearm, to project a nail or other fastener into concrete or metal.

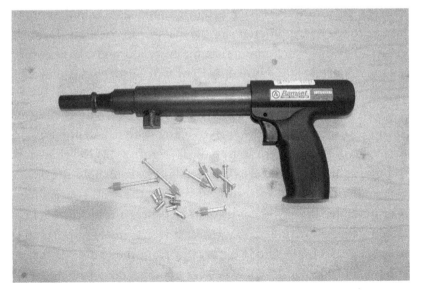

Figure 8-13 Powder actuated gun with nails and loads (charges). *Courtesy Delmar/Cengage Learning*

Before using a powder-actuated tool, a worker is required to have training in its proper use. Some manufacturers offer classes and award a certification card upon successful completion. Before using any powder-actuated tool, it should be inspected for damage and the barrel checked for obstructions. If any damage is found, it must be marked "do not use" and taken out of service for replacement or repair. Always post signs and let other people in the area know that you are going to be using a powder-actuated device. This will prevent them from being alarmed or startled when it goes off, and warns them to use appropriate PPE.

Powder-actuated tools will create flying debris as well as loud noises. Safety glasses are a must for the operator as well as nearby workers, as nail ricochets and pieces thrown from the impact will create flying debris. Powder-actuated tools use a charge which sounds like a gun going off when it is triggered. The noise is extremely loud, so ear protection must be worn by the operator as well as all workers in the area.

When loading a powder-actuated tool, always install the nail or fastener first, followed by the load or charge. If the charge is installed first, there is a possibility that the gun will go off as the nail is being inserted. There are different strengths of charges used with these tools; be sure to use one appropriate for the material in question. The most powerful charge is not appropriate for every job, and charges which are too powerful could cause the nail or fastener to go clean through the surface being worked on.

Two actions are required to fire powder-actuated tools. The tool must be pushed against the surface being fastened, depressing the barrel, before the trigger can be pulled. (Figure 8-14) Never attempt to bypass this safety feature. When using the tool, hold the device perpendicular to the work surface and use the spall guard whenever possible. This will help prevent ricochets and thrown debris. When using powder-actuated tools, it is extremely important to keep hands and other body parts away from the end of the barrel and the location being nailed. Be sure to have a good footing, never overextend, and keep a secure grip on the tool to prevent slips, falls, and loss of control.

Figure 8-14 Powder actuated gun fastening plywood to a concrete wall.
Courtesy Delmar/Cengage Learning

Never carry the fastening devices or other hard objects in the same container as the charges. If the firing cap is struck, they could become lethal. Always store and carry the tool unloaded, and never load the tool until immediately before use. (Figure 8-15)

Ladders

Ladders are one of the most frequently-used tools, while at the same time one of the most misused. Understanding and following the rules for ladder safety will help reduce falls, which are the leading cause of death on construction sites. Reading the safety labels on the ladder will provide information specific to the particular ladder you are using. (Figure 8-16)

As with all tools, a ladder must be thoroughly inspected before use. If any damage is found, the ladder must be marked "do not use" and taken out of service immediately. Examples of damage can include missing parts, broken rungs, broken feet, cracked or splintered rails,

Figure 8-15 Nails and loads should never be stored together in the same container to avoid accidental discharge. *Courtesy Delmar/Cengage Learning*

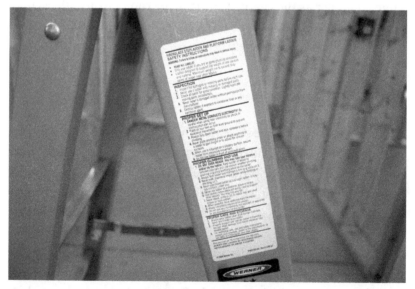

Figure 8-16 Ladders will have instructions for proper care and use on the side of the ladder. *Courtesy Delmar/Cengage Learning*

Figure 8-17 This ladder has a damaged foot. *Courtesy Delmar/Cengage Learning*

and broken spreaders. (Figure 8-17) A loose or wobbly ladder will sway and move while the worker is on it. This can be extremely dangerous and cause the worker to lose his or her balance. Loose and wobbly ladders must also be taken out of service immediately.

When climbing a ladder, a worker should always face the ladder and maintain three points of contact. The three points could be two feet and one hand or two hands and one foot. Having three points of contact helps prevent a fall if a foot or hand slips, as there will still be two contact points. It isn't possible to maintain three points of contact while carrying tools up the ladder, which means that all tools and materials that need to go up or come down with the worker will have to either be in a tool pouch or be hoisted with a towline. Never overextend or lean over while on a ladder, as this can cause you to lose your balance and fall, or cause the ladder to tip over. Never climb a ladder taking more than one step at a time, jump down from a few steps up, or slide down the ladder by the rails.

Most ladders are only rated for one person. The movement of an additional person on the ladder can cause the other person to lose their balance, and will likely overload the ladder.

Never use a ladder if you are tired or dizzy, as you will be more likely to slip and fall. Always wear slip-resistant shoes or boots and make sure they are clean. Trying to climb a ladder with boots full of mud or snow will create a slip hazard. Never use a ladder on uneven ground, or try to prop one leg with a board or other material. Make sure that all the legs have secure footing to prevent the ladder from shifting or falling over once you climb on. Do not use ladders in high wind situations or severe storms.

Aluminum and conductive ladders should never be used near electricity. A conductive ladder which is connected to ground will provide a grounding path which could lead to an electrical shock in the case of a live electric circuit.

Stepladders

When using stepladders, the base must be completely spread apart with the spreaders locked into place. (Figure 8-18) Not having the spreaders locked into place will cause the ladder to be more likely to shift, or it may fold up and tip over. A stepladder should not be climbed while it is leaned against a wall. Stepladders are not designed to be used that way and are likely to slip, causing the worker to fall. Be sure that the four feet are placed on level, secure ground. If the ground is soft or uneven, the ladder may shift once the worker climbs on, causing the worker to lose balance or the ladder to fall over.

Always set up a stepladder as close to the work area as possible to avoid leaning and overextending. If the ladder needs to be moved, always get off before shifting it. Trying to hop the ladder over or hanging on to something and attempting to drag it can lead to falls.

The top cap and top step of stepladders are not to be stood on. The ladder will have warning signs indicating the highest step that can be used; it is extremely easy to lose your balance if you ascend above the maximum level. Most ladders will have steps on the front for climbing and horizontal bars on the back side, which are not intended as steps. Never climb these bars, as they are designed to strengthen the ladder, not support a person.

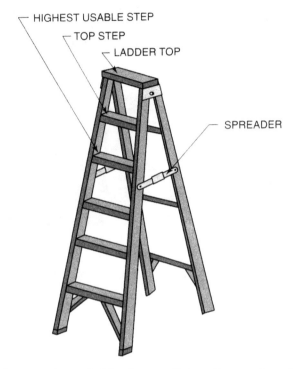

HIGHEST USABLE STEP
TOP STEP
LADDER TOP
SPREADER

Figure 8-18 Stepladder. *Courtesy Delmar/Cengage Learning*

Many ladders have a top cap with various-sized holes and trays for holding tools and materials. (Figure 8-19) Although this is a handy feature, it is important to always remove all tools and materials from the top of the ladder before attempting to move it. Anything left on top of the ladder is likely to fall and land on your head.

Extension Ladders

Extension ladders must be set properly to prevent them from tipping over. (Figure 8-20) The two feet must be on level, secure ground. Uneven ground will cause the ladder to lean; soft ground will give way, causing the ladder to tip; and loose or slippery ground will cause a ladder to slip. Any of these situations are extremely danger-ous and could lead to a fall.

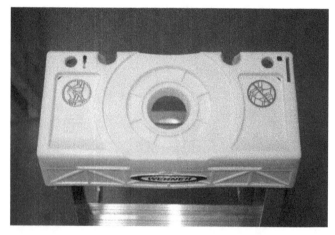

Figure 8-19 Many stepladders are built to hold some tools and materials. All items must be removed from the top of the ladder before it is moved to avoid injuries from falling objects. *Courtesy Delmar/Cengage Learning*

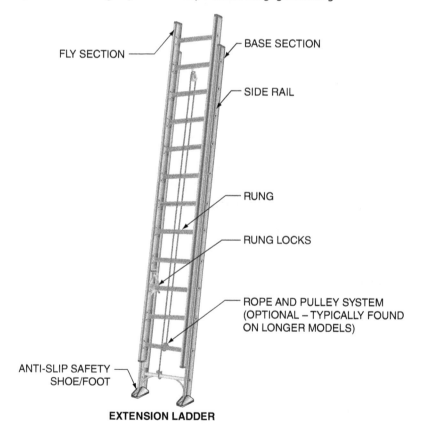

EXTENSION LADDER

Figure 8-20 Extension Ladder. *Courtesy Delmar/Cengage Learning*

Extension ladders should be erected at an angle of 75½ degrees. A good rule of thumb to achieve this angle is to use a four-to-one ratio. For every four-foot length of ladder, the base of the ladder should be one foot from the wall, so a 20-foot long extension ladder should have its base five feet from the wall. (Figure 8-21)

If an extension ladder is to be used to access a roof or other platform, the ladder must extend at least three feet above the platform, and must be secured to the platform before a worker attempts to get on or off.

Scaffolds

When ascending or descending a scaffold, or a ladder to reach a scaffold, extreme care must be taken. (Figure 8-22) Ladders used to access a scaffold must extend three feet above the platform and be secured to it. If a scaffold is ten feet or more in height, fall protection must be used. Fall protection can be in the form of guardrails, or a body

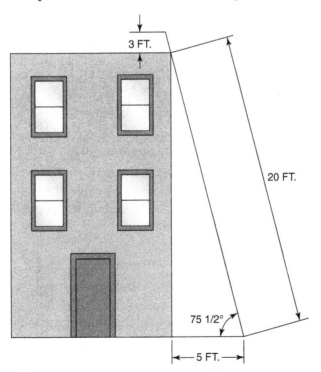

Figure 8-21 Extension ladders should be set at a 75 ½ degree angle. That means that for every 4 feet of ladder length, the base must be 1 foot away from the building. *Courtesy Delmar/Cengage Learning*

Figure 8-22 Scaffold. *Courtesy Delmar/Cengage Learning*

harness with a lanyard attached to a secure anchor point. Most scaffold manufacturers advise against tying off to the scaffold due to the forces involved with arresting a fall. When walking on a platform, be aware of your footing, as planks may shift and there could be trip hazards.

When working on, below, or near a scaffold, safety glasses and a hard hat are required. There will often be dropped objects which could cause serious injuries. When working on a scaffold, be sure to secure all tools and materials to prevent injuring workers below.

Aerial Lift

Aerial lifts are used to get workers up to an elevated level to perform a task. The most common aerial lifts used by electricians are scissor lifts, boom lifts, and bucket trucks. Aerial lift accidents account for an average of 26 construction deaths yearly. The majority of the injuries and deaths in aerial lifts are from electrocutions, falls, and tipovers. It is imperative that workers who are going to use an aerial lift have the necessary training on the proper use of the machine. The manufacturer of the lift will have a list of items which must be checked before every use.

Boom Lifts and Bucket Trucks

An average of 18 construction deaths yearly result from accidents involving boom lifts (Figure 8-23) and bucket trucks.(Figure 8-24) Approximately one-half of these deaths are from electrocution. When using a lift, the worker must keep a very close eye on any overhead lines which may be nearby, and maintain a safe distance. Always disconnect the power to equipment which is to be serviced and use the appropriate steps to verify the absence of voltage.

When entering the bucket, be sure to latch any safety chains or doors which will help prevent falling out. Attach your safety harness to the appropriate location using a lanyard or positioning device. Always stand on the floor of the bucket or platform and never climb up onto the rail to gain extra height, or lean over the rail. Never try to gain extra height by using a ladder in a bucket or on a boom lift platform.

Care must be taken when operating a lift, as some lift controls are very sensitive and may cause jerky movements. This could cause a part of the body to get pinned between the bucket and another object, or eject the worker.

Figure 8-23 Boom Lift. *Courtesy Delmar/Cengage Learning*

Figure 8-24 Bucket Truck. (The Versalift product line is manufactured by Time Manufacturing Company in Waco, TX.) *Courtesy Delmar/Cengage Learning*

Always use outriggers when necessary. (Figure 8-25) The necessity of outriggers will be determined by the type of lift and the situation; it may also be necessary to set the brake and chock the wheels.

Some models of lifts allow the worker to move the lift while the worker is elevated, while others will not. If moving the lift is permitted while the worker elevated, extreme caution must be taken, as a very small drop or bump for a tire may swing the boom a large distance. Watch out for large obstacles, holes and drop offs, and do not drive onto soft or uneven surfaces while the lift is elevated,as this can cause a slingshot effect and eject the worker or cause serious injuries. Do not exceed the manufacturer's load limits, as this may cause a tipover.

Scissor Lifts

An average of eight construction deaths yearly are from accidents involving scissor lifts. (Figure 8-26) About one-half of these are from falls. Always latch the safety chains and close the doors after entering the platform. Always stand on the floor of the platform; never climb up onto the rail to gain extra height, or lean over the rail. Never try to gain extra height by using a ladder on the lift platform. While the lift is elevated, never attempt to climb up or down the scissors to get on or off the lift. The only safe way to get on or off the lift platform is while it is in the down position.

Always be aware of overhead electrical lines and be sure to maintain a safe distance. When working on electrical equipment, be sure to disconnect the power and use the proper procedure to verify the absence of voltage.

Figure 8-25 Outriggers are used to stabilize the lift or bucket truck and prevent a tipover. (The Versalift product line is manufactured by Time Manufacturing Company in Waco, TX.) *Courtesy Delmar/Cengage Learning*

Some scissor lifts are designed for rough surfaces, while others are only for use on smooth surfaces. Never take a lift designed for smooth surfaces onto a rough surface, as it may tip over. Watch out for large obstacles, holes, or drop-offs when driving a scissor lift, particularly while it is elevated. A sudden drop by one tire will create a slingshot effect on the worker up on the platform.

Machinery Safety

Machinery used by electricians and other construction workers includes trenchers, skid steer loaders, backhoes, etc., (Figure 8-27) Machine safety starts with proper training. Workers must be trained

Figure 8-26 A scissor lift will raise straight up giving an elevated platform to work from. *Courtesy Delmar/Cengage Learning*

Figure 8-27 Trencher. *Courtesy Delmar/Cengage Learning*

on the specific machine that is to be used: if a different model is to be employed, the worker must be trained on the proper use of that particular model. It takes many hours of use for a person to become comfortable and proficient on a piece of equipment. When operating a piece of machinery, care must be taken to not injure workers that are nearby. If you are working near machinery always be aware of where it is, where it could go, and keep a safe distance. They move quickly and have many blind spots.

Before using a trencher, skid steer, etc. it must be inspected for damage and disrepair. The manufacturer will have a list of items which must be reviewed before each use. If there is any damage or other issues the machine must be taken out of service.

Before trenching, digging, drilling holes, etc., all underground utilities must be located. Failure to have underground utilities marked before digging can result in civil penalties, as it can be extremely dangerous and expensive if they are accidentally hit or cut. Most parts of the country have a hotline phone number which can be dialed to have all the underground utilities marked. Examples of underground lines that will be marked include electric, telephone, cable TV, natural gas, fiber optic, water, and sewer. Keep in mind that these companies will only mark the utility lines; any underground lines which are owned by the customer will usually not be marked. Examples of underground lines that will not be marked include underground feeds to separate buildings, parking lot lights, signs, etc.

Always wear safety glasses and a hard hat. When operating or working near machinery there will be debris being thrown around, which can cause injuries. Machines such as trenchers, skid steer loaders, backhoes, etc. are very noisy. While operating them or working nearby, hearing protection is a must. Wear the appropriate clothing: loose baggy clothing can get caught in rotating parts. Trenchers and other machines have lots of moving parts, so be sure to keep your head, hands, and feet clear.

Summary

- Safety glasses must be worn at all times when working with tools and equipment.
- Always disconnect the power before working on a circuit.
- Before using power tools, a person must receive training on their proper use and safety procedures.
- Only use tools for their intended task, as misuse can lead to injuries.
- Damaged tools and equipment must be removed from service immediately.
- Never bypass safety features on tools and equipment.
- Always ensure ladders have a secure footing before attempting to climb.
- Always climb facing the ladder.
- When using an aerial lift, always look for energized power lines.
- Always keep your feet on the floor of an aerial lift and use the appropriate fall protection.

Review Questions

1. What should always be done before working on an electrical circuit?
2. What two items of personal protective equipment will often be necessary when working with power tools?
3. Why should loose clothing and jewelry be avoided?
4. What types of tools have a two-wire power cord?
5. What procedure should be followed if damage is found on a tool or piece of equipment?
6. What type of ladder should never be used around electricity?
7. An extension ladder must extend _____ above a platform that it is accessing, and be _____ to the platform.

8. What two items of personal protective equipment will be necessary when working on or below a scaffold?

9. Why should a worker wear safety glasses and a hard hat on a trencher?

9

Hazardous Working Environments

Objectives

- Recognize the symptoms of heat exhaustion and heat stroke;

- Describe how to prevent heat-related injuries;

- Recognize the symptoms of frostbite and hypothermia;

- Describe how to prevent cold temperature-related injuries;

- Describe the hazards associated with confined spaces.

Introduction

Construction workers often work in environments that can be dangerous, whether it is a one-hundred-plus degree day in the summer or a winter's day with sub-zero temperatures. In addition to the dangers associated with temperature, work locations such as confined spaces may also create a dangerous environment. Workers must be able to recognize these potential hazards and take the necessary steps to avoid a dangerous situation.

Hot Environments

Heat Exhaustion

Construction workers are very susceptible to heat exhaustion due to the nature of their work. Performing strenuous tasks while in the hot sun or a hot environment can cause a person to become very sick, and left untreated heat exhaustion can lead to death.

Symptoms of heat exhaustion:

- Sweating;
- Headache;
- Dizziness;
- Weakness;
- Mood changes (irritability or confusion);
- Upset stomach or vomiting;
- Muscle cramps;
- Fainting.

Workers need to be aware of the symptoms and monitor themselves as well as other workers. If you suspect that you or another worker is suffering from heat exhaustion, the appropriate First Aid procedures must be administered immediately.

Care for heat exhaustion:

- Move the victim to a cool place;
- Provide cool liquids (such as water or drinks with electrolytes);
- Remove excess clothing;
- Moisten and fan the victim.

If the person's condition doesn't improve within 30 minutes, medical care is required. Heat exhaustion can quickly escalate to heat stroke, which if left untreated will be fatal.

Heat Stroke

Heat stroke is a very serious situation and a medical emergency. If you suspect a person is suffering from heat stroke call 911 immediately; left untreated, it will be fatal. Healthy people are not immune to heat stroke; strenuous work in hot environments can lead to a heat stroke for anyone.

Symptoms of heat stroke:

- Hot skin, either dry or wet (Many heat stroke victims won't sweat and will have dry skin, but this is not always the case);
- Severe headache;
- Mental confusion, agitation, or disorientation;
- Unresponsiveness;
- Seizures or convulsions.

The first thing to do if a person is having a heat stroke is call 911. While waiting for help to arrive, First Aid should be administered.

Care for a heat stroke victim while waiting for help to arrive:

- Move the victim to a cool place;
- Cool the victim by any means necessary:
  - Cold water bath;
  - Cool wet towels or sheets;
  - Fans;
  - Cold packs in armpits, neck, and groin.
- Elevate the head and shoulders slightly.

Warm Weather Injury Prevention

It is important to prevent heat-related injuries rather than have to treat them. To help prevent heat exhaustion and heat stroke when working in hot environments, the following should be kept in mind:

- Perform the most strenuous work during the coolest part of the day;
- Wear light-colored clothing that is loose-fitting and breathable;

- Drink small amounts of water frequently;
- Avoid drinks with caffeine;
- Eat small meals;
- Take frequent breaks in a cool or shaded area;
- Work in the shade if possible;
- Pay attention to your body.

Sun

A lot of construction is performed outside under the effects of the sun. Given the amount of exposure construction workers have to the sun, they must take proper precautions to prevent skin damage. Sunburns and suntans are a visual indicator that the skin has been damaged.

A severe sunburn will make life at work extremely uncomfortable and will make it difficult to perform many tasks. Although a sunburn will heal in a few days, there are some very serious long-term effects from sun damage, such as cataracts, premature aging of the skin, wrinkles, and skin cancer.

Sunglasses or safety glasses with UV protection will help to protect your eyes from sun damage, while wearing a wide-brimmed hat will shade the head, face, and ears. Always cover your skin with clothing or sunscreen. Sunscreen should have an SPF (Sun Protection Factor) of at least 30 and be applied according to the sunscreen manufacturer's recommendations.

Cold Work Environments

During the cold-weather months, construction workers will often be working in cold environments, whether outside or in buildings which aren't heated yet. Taking the proper precautions will make working in cold environments safer and much more comfortable. (Figure 9-1)

Figure 9-1 Working out in the cold can be dangerous if the proper precautions aren't taken. *Courtesy Delmar/ Cengage Learning*

Frostbite

Frostbite occurs when tissue is damaged due to freezing. The temperature of the outside air will determine how quickly frostbite will set in: the colder the temperature, the quicker damage will occur. Wind also plays an important role, as it will intensify the effects of the cold. On a cold, windy day frostbite can happen after just a few minutes. (Figure 9-2)

Frostbite typically affects the ears, nose, cheeks, chin, fingers, and toes. Severe frostbite can cause permanent damage, resulting in gangrene and/or amputation. It is very difficult to recognize frostbite on yourself unless you are looking in a mirror. Keep a lookout for frostbite on fellow workers and have them keep an eye on you. If frostbite is noticed, it must be cared for as soon as possible. Severe frostbite will require medical attention.

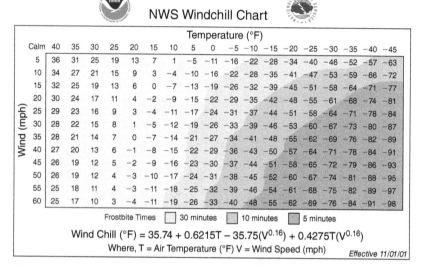

NWS Windchill Chart

Temperature (°F)																	
Calm	40	35	30	25	20	15	10	5	0	−5	−10	−15	−20	−25	−30	−35	−40	−45
5	36	31	25	19	13	7	1	−5	−11	−16	−22	−28	−34	−40	−46	−52	−57	−63
10	34	27	21	15	9	3	−4	−10	−16	−22	−28	−35	−41	−47	−53	−59	−66	−72
15	32	25	19	13	6	0	−7	−13	−19	−26	−32	−39	−45	−51	−58	−64	−71	−77
20	30	24	17	11	4	−2	−9	−15	−22	−29	−35	−42	−48	−55	−61	−68	−74	−81
25	29	23	16	9	3	−4	−11	−17	−24	−31	−37	−44	−51	−58	−64	−71	−78	−84
30	28	22	15	8	1	−5	−12	−19	−26	−33	−39	−46	−53	−60	−67	−73	−80	−87
35	28	21	14	7	0	−7	−14	−21	−27	−34	−41	−48	−55	−62	−69	−76	−82	−89
40	27	20	13	6	−1	−8	−15	−22	−29	−36	−43	−50	−57	−64	−71	−78	−84	−91
45	26	19	12	5	−2	−9	−16	−23	−30	−37	−44	−51	−58	−65	−72	−79	−86	−93
50	26	19	12	4	−3	−10	−17	−24	−31	−38	−45	−52	−60	−67	−74	−81	−88	−95
55	25	18	11	4	−3	−11	−18	−25	−32	−39	−46	−54	−61	−68	−75	−82	−89	−97
60	25	17	10	3	−4	−11	−19	−26	−33	−40	−48	−55	−62	−69	−76	−84	−91	−98

Wind (mph)

Frostbite Times ☐ 30 minutes ☐ 10 minutes ☐ 5 minutes

$$\text{Wind Chill (°F)} = 35.74 + 0.6215T - 35.75(V^{0.16}) + 0.4275T(V^{0.16})$$

Where, T = Air Temperature (°F) V = Wind Speed (mph)

Effective 11/01/01

Figure 9-2 This chart gives the approximate time it takes to receive frostbite based on the temperature and wind speed. *Courtesy National Weather Service website*

Symptoms of frostbite:

- Skin color is white, grayish yellow, or waxy;
- Skin is numb;
- Skin is cold and firm;
- Area which is painfully cold suddenly stops hurting.

Care for frostbite:

- Move into a warm environment (if possible, do not walk on frost-bitten feet or toes, as this will increase the damage);
- Remove clothing or other items (rings, jewelry, boots) which could impair blood flow to frostbitten area;
- Do not rub or massage frostbitten area, as this could cause more damage;
- Immerse the affected area in warm water or use body heat to warm the area. A warm, wet cloth can be used for areas on the face. Be careful with the temperature of the water used to warm the frostbitten area: the body will be numb and will not be able to sense scalding;

- Do not thaw a frostbitten area out and then allow it to refreeze, as this will cause severe damage;
- Seek medical attention.

Frostbite can be prevented by covering up any exposed skin and dressing appropriately to keep the body warm. Examples of items that should be worn to protect the areas of the body sensitive to frostbite include warm socks with insulated boots, insulated gloves, a facemask, and a hat. Disposable warming packs are available to help warm up areas that tend to get cold, like fingers.

Hypothermia

Hypothermia is a condition where the body is losing heat faster than it can produce it. When a person's core body temperature has dropped below 95 degrees Fahrenheit they are considered hypothermic. Hypothermia will alter a person's mental state, so they will often not be able to recognize that it is happening. It is extremely important to keep an eye on fellow workers; if you suspect a person is becoming hypothermic, they must be treated immediately. Hypothermia is a medical emergency which requires immediate medical attention. Once hypothermia sets in, the body may not be able to recover on its own.

Although working in extreme cold can cause hypothermia, it can also happen in cool temperatures, particularly if a person is wet from rain, sweat, etc. Care must be taken not only in cold environments, but in cool ones as well.

Symptoms:

- Change in mental state (becoming disoriented, confused, drowsy, apathetic, or aggressive);
- Fumbling hands;
- Mumbling and slurred speech;
- Vigorous shivering (if the body temperature drops below 90 degrees Fahrenheit, shivering will stop);
- Cool abdomen, chest, or back;
- Low core body temperature;

- Slowed breathing;
- Unconsciousness.

To care for a person who is suffering from hypothermia, you must use the proper First Aid procedures and call 911 for professional medical help. The 911 operator can provide information on what to do for the victim until help arrives.

Care for hypothermia while waiting for help to arrive:

- Move them into a warm area;
- Replace wet clothing with dry clothing and blankets or other layers;
- If the person is shivering, do not attempt to stop it by applying heat: shivering is how the body heats itself up;
- If the body is not shivering, and you are in a remote area where emergency help will be a while coming, the body can be warmed by using skin to skin contact;
- Provide warm beverages (not alcohol).

Cold Weather Injury Prevention

It is important to prevent cold weather-related injuries. By understanding the dangers, being smart about the working environment, dressing appropriately, and recognizing the initial symptoms, a person can safely work in cold environments.

Pay attention to the weather forecast and the type of work that is to be performed. Knowing if the temperatures will be sub-zero, if it is going to snow or rain, or if it is going to be windy will determine how you will need to dress. Dressing appropriately will keep you comfortable and safe.

When working in cold environments, always dress in layers; a base layer of long underwear, middle layers that can be added and removed as needed, and an outer shell that is waterproof and breathable to allow moisture to escape.

It is important to wear fabrics that will work for you rather than against you. Cotton is not a good material to wear in cold temperatures. Cotton tends to retain moisture and keep it close to the body; eventually the moisture will cool off and cause the wearer to become cold. Fabrics made out of polypropylene, silk, fleece, and wool tend to work better, as they wick moisture away from the body while retaining body heat. (Figure 9-3) Polypropylene and silk work well as a base layer, while fleece and wool work well as the middle layers that can be added and removed as needed. Although synthetic fibers work very well in cold-weather situations, remember that when working in an area where there is an arc flash potential, synthetic fibers can be extremely dangerous as they will burn or melt.

It is extremely important to keep dry when working in cold environments. If you get too warm and begin to sweat, the moisture

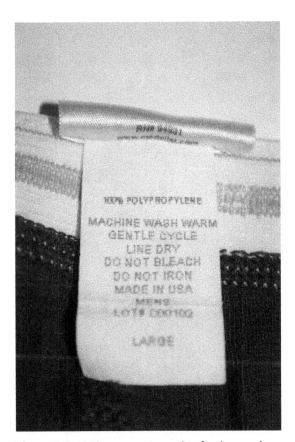

Figure 9-3 Undergarments made of polypropylene wick moisture away from the body. *Courtesy
Delmar/Cengage Learning*

in your clothing will cool off, eventually making you cold. The key is to prevent sweating by removing layers. Keeping the appropriate layers of clothing on without getting too cold or too warm can be a tricky task, as the intensity of work constantly changes throughout the day.

Insulated bibs or coveralls are popular choices for outer shells in the construction industry. (Figure 9-4) The outer material is a tough fabric that will hold up well to the abuse of day-to-day activities. The drawback is that they are often not waterproof, so if they are subjected to wet conditions they may become saturated. Consideration must also be given to the fact that most are not flame-resistant, so they should not be used where an arc flash potential exists.

Don't forget about the head, hands, and feet. Much of a person's body heat escapes through the head. A hat and facemask will prevent some of this heat loss and make a person much more comfortable. The hands are a difficult part of the body to keep warm when working in the winter. Choose gloves that are as warm as possible

Figure 9-4 Insulated Bibs. *Courtesy of Carhartt*

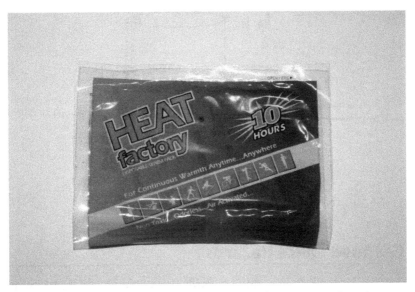

Figure 9-5 Disposable hand warmer. *Courtesy Delmar/Cengage Learning*

while still providing the required dexterity. Having hand-warmers along can give the fingers an occasional boost of warmth and will fit into some gloves. (Figure 9-5) The feet are another critical area which, when cold, can make life miserable. Warm socks with good insulated boots can help avoid this problem. Just as with other cold-weather gear, care must be taken to prevent the feet from getting too warm, which will cause them to sweat; if a person's feet are wet they will quickly become cold. This can be difficult to control if a person is working in both warm and cold environments throughout the day.

Confined Spaces

Electricians will occasionally find themselves working in confined spaces. A confined space is an area with limited entrances and exits, not intended for people to work in continuously, but where a worker may be required to perform a task. Examples of confined spaces include tanks, manholes, sewers, silos, tunnels, vaults, trenches, pits, and so on. (Figure 9-6)

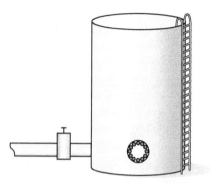

Figure 9-6 Tanks are a confined space.
Courtesy Delmar/Cengage Learning

Confined Space Hazards

Confined spaces often contain hazards that could be life-threatening. When a confined space contains a life-threatening hazard, it becomes a permit-required confined space. Examples of hazards found in confined spaces include but are not limited to: low or high oxygen levels; toxic or explosive gases, fumes, or materials; electrical hazards; and crushing or engulfing hazards. Workers must have the necessary training and follow the appropriate procedures before attempting an entry.

Low or High Oxygen Levels

Low oxygen levels can cause a person to become unconscious and could lead to death. Varying oxygen levels can be found in almost any confined space. Low oxygen levels can result from the process of corrosion, rotting of vegetable matter, combustion from gas engines, welding, and simply breathing. High oxygen levels can cause a person's thinking to become impaired and can be explosive. Before entering, an air sample must be taken to check the levels. It may be possible to use ventilation to purge the space and to supply fresh air while the work is being done. Other times, it may be necessary to use a full face respirator to supply breathable air. The air must not only be sampled before the work commences to ensure that levels are safe, but also as the work is being performed, to make sure they haven't changed.

Toxic or Combustible Atmospheres

Toxic gases or vapors can quickly cause asphyxiation and can be found in almost any confined space. Combustible gases or vapors can easily become ignited by arcs resulting from the use of hand tools, power tools, lamps, etc. Before entering, an air sample must be taken to check the levels. Ventilation may be used to purge the space, but it may not be able to eliminate the hazard. A full face respirator that supplies breathable air and tools designed for hazardous locations may be required.

Electrical Hazards

Electrical hazards that may be present in confined spaces include electrical shock and arc flash. Manholes, electrical vaults, and many other confined spaces may contain these hazards. Disconnecting the power and using the appropriate lockout-tagout procedures should be used to eliminate the hazard. In the event that power cannot be removed, the appropriate personal protective equipment must be used.

Crushing or Engulfing Hazard

Trenches with sufficient depth become a confined space, as they have the potential to cave in, engulfing a person. If the trench is deep enough to have a cave-in hazard, it must be shored using appropriate methods. Shoring the trench will protect the person working in the event of a cave-in. Silos and bins which contain granular materials also pose an engulfing hazard. A retrieval system must be in place in the event of an accident.

Confined Space Entry

There must always be at least two people directly involved in confined space work: a worker to enter the space and an attendant to monitor. The worker must have training on performing confined space work, be aware of all the potential hazards, and wear the appropriate PPE. The attendant must have training on confined space work, be aware of all the potential hazards, monitor the working conditions, and know how to retrieve the worker in the case of an incident. The attendant must never go into the confined space to help the worker—for any reason. If the worker becomes unconscious and the attendant goes in to help, it is very likely that the attendant will also be overcome by the same conditions, resulting in two dead workers. A retrieval system must be used to remove the worker in the case of an accident. A retrieval system consists of a harness worn by the worker and a retrieval device to pull the worker out of the confined space. (Figure 9-7)

It is extremely important that all workers follow the confined entry safety procedures, as failure to do so may lead to severe injury or death. The necessary procedures will vary depending on the space and will be detailed by the employer's confined space entry program.

Figure 9-7 Confined space retrieval system. *Photo courtesy of Miller® Fall Protection*

Summary

- Heat exhaustion can quickly lead to heat stroke and must be cared for immediately.
- Heat stroke is a life-threatening situation that must be immediately treated by a medical professional.
- To prevent heat-related injuries when working in hot environments, workers must wear the appropriate clothing, keep well hydrated, and use warm-weather work practices.
- Hypothermia is a life-threatening situation that must be immediately treated by a medical professional.
- To prevent cold temperature-related injuries while working in cold environments, workers must wear the appropriate clothing.
- Confined spaces can potentially contain many types of hazards. Workers must have the appropriate confined space training and follow all safety requirements to prevent injuries that could result in death.

Review Questions

1. List five symptoms of heat exhaustion.
2. List five symptoms of heat stroke.
3. List five things that can be done to help prevent heat-related injuries.
4. What can be done to protect a person from sun damage?
5. List four symptoms of frostbite.
6. List five symptoms of hypothermia.
7. What type of fabric should be avoided for cold-weather wear?
8. List four hazards associated with confined spaces.

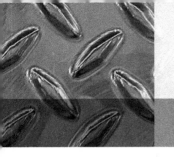

Index

Printed in the USA
CPSIA information can be obtained
at www.ICGtesting.com
JSHW012332230924
70181JS00002B/21